The Global Canopy Handbook

Techniques of Access and Study in the Forest Roof

Published
with the generous support of

Rufford
The Foundation

Printed
on recycled "Revive" paper using vegetable based inks

Robert Horne Group

Edited by
Andrew W. Mitchell
Katherine Secoy
Tobias Jackson

Publisher

GCP
Global Canopy Programme

Front Cover:
Laurent Pyot hangs from the
'Canopy Bubble', a helium filled
balloon, to gain access to the
forest roof above the coast of
Cap Masoala, Madagascar, 2001.
Photo: courtesy of Laurent Pyot

Published by

Global Canopy Programme
Halifax House
Oxford University
South Parks Road,
Oxford OX1 3UB, UK
www.globalcanopy.org

© Global Canopy Foundation 2002
ISBN 0-9542970-0-8
Printed in United Kingdom.

Acknowledgements

First, I must thank the Rufford Foundation for its vision in supporting the publication of this book and the development of the Global Canopy Programme. Secondly, the many authors and photographers all over the world who took up the call to contribute to this book in a short time, for little or no reward, deserve my special thanks. Without their enthusiasm it would have been impossible. I would also like to thank Katherine Secoy and Tobias Jackson who helped to compile and edit this publication. My thanks also go to Jennie Bayliss and the team at Impressions Design who kindly designed and typeset the book in record time and Mandy Klyne for proof reading. Finally my thanks go to Robert Horne Paper Group for their generous support for this publication.

Dedication

This book is dedicated the late Alan Smith who first had the inspired idea of using a large jib crane to explore the rainforest roof.

Alan Smith (right) and Andrew Mitchell
aboard the first canopy crane
in Panama, c. 1990.

contents

PART 1
Methods of Access in Forest Canopies

PART 2

Methods of Study in Forest Canopies

GCP
Global Canopy Programme

A global alliance linking studies of forest canopies worldwide into a collaborative programme of research, education and conservation addressing biodiversity, climate change and poverty alleviation.

This book was published on the official launch of the Global Canopy Programme, at the Third International Canopy Conference in Cairns, Australia 28th June 2002.

TWO of the greatest challenges facing our natural world are the loss of biological diversity and the impact of climate change. The threats presented by these issues affect every nation, and nowhere are more acute than in the world's great forests. One part of these forests could hold a vital key to understanding the risks these threats pose to people and business; yet is still the least known terrestrial habitat on earth. This is the high canopy of the world's forests. The reasons for its significance are simple. Firstly, almost half of all the earth's biodiversity may exist in forest canopies, much of it undiscovered and in the tropics. Secondly, the canopy interface with the atmosphere is poorly understood yet it directly effects climate change and the carbon balance of the earth. As forest destruction proceeds apace, the window of opportunity to document the value of the life it contains or its potentially crucial role in maintaining the stability of our planet is closing rapidly.

The Global Canopy Programme (GCP) seeks to address this issue in a significant and far reaching way. Many researchers using cranes, balloons, walkways, towers and climbing ropes are now collaborating to create the first co-ordinated effort to investigate this critical environment on a global scale. Considerable knowledge of value to civil society, enhanced benefits for local communities, as well as critical information needed by policy makers, will be the outcome. Following our intervention, The Convention on Biological Diversity – Expanded Programme of Work on Forest Biological Diversity – adopted at COP 6 in April 2002, now calls on Governments to place a greater emphasis on the forest canopy interface with the atmosphere and the endangered species it contains. Through the GCP, we aim to respond specifically to this call and we hope the Global Canopy Handbook will provide an added stimulus to the process. Details of the GCP can be found in Section 12 of this book and those interested in supporting it or contributing projects to the programme, are invited to visit our website or to contact the GCP Secretariat.

Andrew Mitchell
Director, Global Canopy Programme

Contact details

The Global Canopy Programme
Halifax House, University of Oxford, 6–8 South Parks Road, Oxford OX1 3UB, UK
Tel: 44 (0) 1865 271036   Fax: 44 (0) 1865 271035
Email: a.mitchell@globalcanopy.org   Web: www.globalcanopy.org

Foreword

President,
International
Canopy Network

I<small>N</small> 1982, I attended a symposium on the nascent field of canopy research at the University of Leeds in England. It was the first international meeting of these "explorers of the high frontier," many of whom have since continued to study forest canopies. One product of that meeting was a slim booklet that has resided on my book shelf ever since – *Reaching the Rain Forest Roof.* I pulled it down again to compare it to the book you hold in hand, also published by Mitchell and his colleagues two decades later.

What does the comparison reveal? First, the canopy researcher of 1982 seemed consumed with methods of getting up to the canopy safely and efficiently. In contrast, a quick glance through the Table of Contents of this 2002 handbook reveals that contemporary canopy researchers are concerned about much more than access. They have developed ways to document the composition, structure, and functional roles of canopy organisms.

Second, early work in the canopy was almost entirely descriptive. Participants in the Leeds symposium showed images of new species and specialized microhabitats never encountered on the forest floor. There is greater emphasis on quantitative work. We now recognize that "standard" statistical methods generated for ground-bound forest research may not be appropriate for canopy questions. The structural complexity of trees requires new applications of topology and statistics.

Third, the field has developed an emphasis on data management. At the Leeds conference, there were so few data that there was no obvious need to formally manage it. In contrast, today's canopy researchers' methods of gathering data – especially via remote sensing and satellite imagery – make it necessary to spend considerable effort on the storage, exchange, and archiving of canopy data, just as in other mature fields of science. This is manifested in the four sections within an entire chapter devoted to data management.

There are still places we have yet to explore. For example, the field of visualization of forest canopy data is not yet developed, and we have no chapter in this handbook to document the ways in which trees can be visually represented. That – and other work – remain for those reading this book to investigate.

The final chapter – on networks related to canopy studies – makes apparent the growing connectivity among canopy researchers and between researchers and policy-makers, educators, and conservationists. Of all the areas of canopy research effort, I think this chapter would be most surprising to our Leedsian predecessors – that in such a short period of time – 20 years, shorter than the lifetime of an oak sapling – canopy researchers have moved from struggling to scale the heights of a forest tree to struggling to communicate with decision-makers, large government agencies and media makers.

It has been an exciting and learningful journey for those who have been involved in canopy studies since the Leeds Conference. It promises to be even more exciting and useful to society in the future. This second handbook on canopy work will ease that journey for those who follow.

Introduction

By
Andrew Mitchell
Director,
Global Canopy
Programme

I N 1982, the Leeds Philosophical and Literary Society published a little booklet entitled *Reaching the Rainforest Roof* (Mitchell 1982) as a guide for those wishing to explore, what at that time, was an almost completely unknown world. Twenty years on, it seemed a good idea to produce a new edition as, since then, much has changed. The extent of undisturbed forest left in the world has shrunk dramatically. At the same time the discoveries made in forest canopies by pioneering researchers such as Lim Boo Liat, Nalini Nadkarni, and Meg Lowman who first journeyed there on climbing ropes and walkways, along with some rather modest efforts by myself, have helped to reveal the extent of our ignorance about life on earth (Mitchell1986). The vast quantities of arthropods which researchers such as Terry Erwin, Stephen Sutton and Roger Kitching have discovered there, most not seen before, has caused a wholesale revision of the numbers of species on earth, up from one million to as many as 50 million, a figure still hotly debated. Species thought to be rare, have been found to be common in the canopy. The prospect of new discoveries a few tens of metres above our heads has proved irresistible and a cohort of rather brave naturalists with a talent for innovation and a head for heights has, over the last twenty years, begun to open up this world to the eyes of science and the public.

Over this time, a whole range of new techniques has been developed for accessing the forest roof from below and above, even from space. This book sets out to offer those wishing to investigate forest canopies, a range of choices from monkeys, to balloons, to satellites. The techniques to choose in the 90 metre old growth forests of the Pacific Northwest or the giant tropical dipterocarps of Asia or the 20 metre mixed woodlands of Europe will vary but most of those in use now are here. The most favoured low-tech method is still a climbing rope, but as I get older the desire to access the canopy this way has been overtaken by a fear of falling and a lack of time spent at the gym. For ageing arbonauts like myself, or those without good climbing skills, the use of a large construction crane set up in the forest, is much less dangerous. A simple 'gondola' lowered to the forest floor provides effortless access to all parts of the crown. We have to thank the late Alan Smith in Panama for this leap of imagination. The 11 such cranes now in place and new ones planned in the coming years, are set to revolutionise our knowledge of this environment. Innovation persists; none more so than the remarkable lighter-than-air

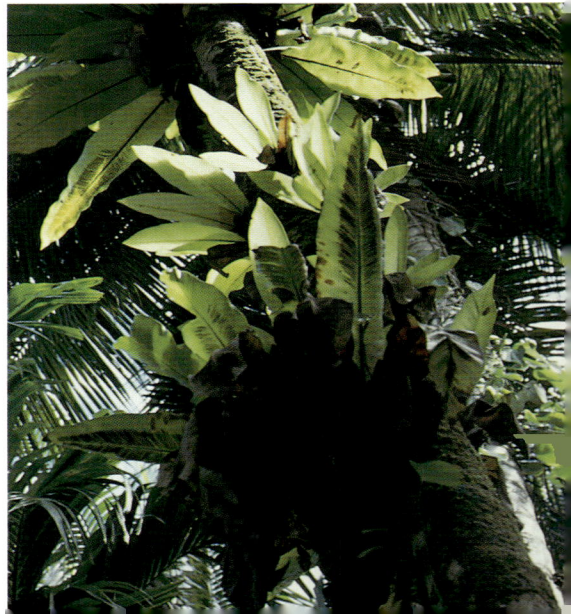

Photo right:
Andrew Mitchell

"Radeau de Cimes" or sky raft, developed by Francis Hallé and his team in France. One-man helium balloons, tethered by a line across the forest roof, now offer low cost access to the outer edge, and the remarkable COPAS tower and cable system due to be installed in French Guiana this year, will enable 1.5 hectares of forest to be reached with relative ease. Remotely sensing the canopy from space or aircraft is perhaps the most comfortable method of all. The improving resolution of these instruments now makes the opportunity of scanning the forest from a computer screen, to determine the composition of the tree species it contains, no longer a remote possibility.

Getting access to the forest roof is merely the beginning. The science that is done there also requires innovation. Assembled here are the latest methods of study that are being used in forest canopies. This is by no means an exhaustive list but it provides an overview of the kinds of work being done in this environment and the techniques being used. Science in the canopy has always been difficult. Hard earned results gained by years of work in one of the most dangerous environments on earth have occasionally been thrown out by the best journals for lack of replication. Perhaps in the future we will see access sites with 3 small cranes operating together, rather than one large one to overcome this problem. Canopy science has had to graduate from a young discipline dominated by questions asking 'what is there' towards those asking 'how does it work'. Another question, 'why does it matter', is now increasingly important as funding for this kind of work becomes linked to relevant social outcomes rather than mere curiosity.

At the Tropical Forest Canopies: Ecology and Management conference in Oxford in 1998, I was impressed by the number of new initiatives in canopy science and exploration but disturbed at the lack of communication and co-ordination among them. Here was the naturalist's 'last biotic frontier' and a major interface between life and the atmosphere, the destruction of which has severe implications for the future of global biodiversity and climate change - yet no major initiative existed to leverage this work into a powerful global programme. I published these concerns (Mitchell 2001), challenged the canopy community to collaborate and called for significant funding to match the task. A year later the plan was laid for a Global Canopy Programme linking research efforts into a global project to investigate the role of forest canopies in maintaining a stable planet (Nadkarni *et al* 2000). The Rufford Foundation and the Maurice Laing Foundation breathed life into these plans and enabled the creation of the co-ordinating Secretariat for the Global Canopy Programme in Oxford in 2001. Now a remarkable alliance of governmental and non-governmental organisations is poised to embark on a ten year investigation of the forest canopies of the world (see section 12). All that is needed is the funding to achieve it. We have high hopes!

Literature cited

Mitchell AW (1982). *Reaching the Rainforest Roof. A Handbook on techniques of access and study in the rainforest roof.* Leeds Philosophical and Literary Society, Leeds, UK. Pp 36

Mitchell AW (1986). *The Enchanted Canopy – secrets from the rainforest roof.* William Collins and Sons, UK. Pp 255

Mitchell AW (2001) Canopy Science – Time to Shape Up. *Plant Ecology* 153: 5-11

Nadkarni N M *et al* (2000). *Forest Canopy Research Planning Workshop Final Report:* Evergreen State College. Unpublished Pp 25.
http://academic.evergreen.edu/individuals/nadkarnn/gcp/report.html

Part 1

Methods of access
in forest canopies

Monkeys, Ropes, Towers and Walkways

Monkeys as Canopy Collectors

By
Valérie
Trichon

THE use of monkeys to collect botanical samples from high in the forest canopy is a technique that has been established for about 15 years on the island of Sumatra, Indonesia. The idea was first used by a team of French botanists working there in cooperation with the Seameo-Biotrop (Bogor, Indonesia). For many years monkeys have been trained by local villagers in Sumatra to climb coconut palms and to retrieve the coconuts and so it seemed possible that this technique could be adapted.

The species of monkey used is a pig-tailed macaque (*Macaca nemestrina* L.) which occurs commonly in this part of the world. The handler attaches a rope to the monkey which is long enough to allow it to climb to the top of the highest trees. Female monkeys are preferred to males, as males can become too big and strong at which point they have a tendency to hold fast on to branches in the canopy and refuse to move if they do not feel like it.

Monkeys were used for this kind of tree sampling because they are very efficient. You have to buy the monkey and to remunerate the handler for taming it and training it to become a good collector, but it is still a cheap investment for the long term. The easiest way to get hold of such a monkey is to try contacting people directly in the villages where they are traditionally trained to fetch coconuts, and explaining your needs.

Research conducted

This collection technique is especially well adapted for intensive botanical inventories over large areas: one monkey was able to sample up to 70 trees per day in Sumatra. The monkey understands a wide variety of orders such as: climb, jump, to the right, to the left, at the end, bite it (the branch), take it, take the fruits, take the flowers, again, that's enough, come down, etc (in Indonesian language, of course).

So far we have only used this technique in Sumatra, but it could theoretically be practised anywhere as long as one can find or bring a monkey and its handler (or perhaps the researcher could learn to handle his own monkey).

Safety

The most important safety consideration is to avoid standing under the tree when the monkey is sampling as dead branches can fall or be thrown down. It also has to

Left: Andrew Mitchell on one of the first canopy walkways, Darien, Panama, 1979. Photo: Scientific Exploration Society.

be remembered that the potential does exist for being bitten by the monkey. People who are to be working closely with the animal should be made aware of this and the necessary precautions taken.

Useful hints

Keep some sweets in your pocket to recompense the monkey during the work, and to attract it in case it doesn't want to come down from the tree. For example, when a tree is fruiting, it is a great opportunity for the botanist but also a great one for the monkey…. just wait and be patient, partially eaten fruits are better than nothing for species identification.

Be careful that the leaves definitely belong to the tree you want to sample, i.e. make sure that you and the handler closely follow the monkey's movement in the canopy. When the animal is tired, she will try to give you any sample within easy reach to save on effort.

Another problem is samples can be damaged during the monkey's activity in the canopy, as the animal is moving very fast in the crowns (specially during descent). In these cases the monkey does not appreciate being made to climb the same tree twice to repeat the work.

Pig-tailed macaques cannot climb thorny or very large trunks. It is however usually possible to overcome this by sending the monkey up a liana or an adjacent tree and then getting it to jump across to the desired one. After rain you need to wait for the bark to be dry enough for the monkey to be able to climb it.

Contact details

Dr Valérie Trichon, Laboratoire d'Ecologie Terrestre, B.P. 4072, 31029 Toulouse cedex 4 - France. Email: trichon@cict.fr

Ropes as a Mechanism for Canopy Access

We have nothing to fear and a great deal to learn from trees.
Marcel Proust, *Pleasure and Regret,* (1896)

By
Martin
Barker
and
Noah
Standridge

Why use rope?

Among the numerous canopy access methods now available, rope techniques continue to be used by many canopy researchers (Nadkarni & Parker 1994, Lowman & Bouricius 1995). In a recent survey of 112 canopy researchers (Barker & Pinard 2001) rope techniques featured strongly and were reported to be the second most utilized method after ground-based techniques. Furthermore, the majority of researchers worked in tropical forests so we can assume that, in many cases, they were working in high canopy forest. Clearly, rope methods are important for canopy researchers.

Rope access is often used in combination with other techniques for reaching the forest canopy. For example, a rope and flexible ladders used together can allow rapid, repeated access to the canopy. Researchers working from the canopy raft (Hallé 1990) have also used rope techniques to increase the vertical range of sampling (Cosson 1995). Conversely ropes can be used for descent only, allowing researchers who use bole climbing techniques (Barker & Sutton 1997) to rapidly descend by abseiling (rappelling) down a rope that has been carried or hoisted up (see below, 'Anchoring the rope').

When are ropes suitable for canopy access?

The suitability of ropes for canopy access depends to a large extent on the type of research that is being conducted. In a survey of canopy researchers, Barker and Pinard (2001) reported that ropes were used most heavily by researchers investigating biodiversity and animal ecology. Ropes may be particularly suitable for these researchers because they provide rapid, flexible and mobile means of access. By comparison researchers working on canopy-atmosphere fluxes or plant ecophysiological studies seem to use ropes less frequently, as they do not provide the long-term and stable support needed for working with bulky or delicate equipment.

Ropes are widely regarded as being of limited use in allowing researchers to reach the outer canopy (Nadkarni & Longino 1990, Longman *et al* 1993, Moffett & Lowman 1995). However, there are methods by which delicate, precarious parts of the canopy can be reached using more stable and reliable supports in the canopy (see below).

Single and double rope technique

There are two main methods to climb trees using ropes; single rope technique (SRT) and arborist technique. The arborist method, developed by professional and recreational tree climbers, involves the climber using a movable rather than fixed rope (Dial & Tobin 1994, Lilly 1994). This approach is particularly effective in climbing very tall trees, or transferring between trees in the upper canopy. The

arborist technique utilizes a relatively short (e.g. 50 m) rope. Both ends of the rope are attached to the climber's harness. One side of the rope provides support and is passed over a stout branch (>15 cm in diameter) that acts as a kind of pulley. The other side of the rope is 'live' and can be pulled though a locking knot or ascender. This allows the climber to ascend the live side of the rope. The arborist technique is particularly suitable for climbing among multiple trees.

In the remainder of this chapter, we will focus on SRT, since this is the most widely used rope climbing method among canopy researchers (see Mitchell 1982, Laman 1995). SRT has its origins in mountaineering and, particularly, caving techniques. In contrast to the arborist technique, SRT generally utilizes one long (e.g. 100 m) rope. The rope first has to be passed over a support branch (see below, 'Getting a rope into the canopy'). One end of the rope is anchored at ground level before the researcher begins to climb up the 'free' side of the rope. SRT is particularly suitable for climbing up and down a single tree.

Although SRT and arborist are distinct techniques, we emphasize that they are not mutually exclusive. For example, an ascent into the lower canopy using a long fixed rope can be followed by movement within tree crowns using a shorter rope (see below, 'Moving off and onto a rope in the canopy').

Rope and related equipment

The two main types of equipment used in SRT are the rope and personal climbing equipment. We will discuss these in turn, focusing on the combination of items of equipment that, in our opinion, work best. Our observations are based on the personal experience of hundreds of climbs, and on the comments of other SRT users. The main items are listed in Table 1, with suggested manufacturers and current (approximate) costs in US$.

Rope

Canopy researchers employing SRT generally use so-called 'static' ropes developed for caving (speleology), which have little stretch. In contrast, the 'dynamic' ropes used in mountaineering and rock climbing are unsuitable and due to their elasticity make the initial ascent on a long rope difficult, and extremely tedious. Descents can also be difficult to control and dangerous as the climber approaches the ground.

SRT ropes are typically constructed from nylon or polyester kernmantle. They have a thick outer sheath, which resists abrasion. The choice of rope diameter depends on intended use. Thicker ropes (e.g. 11 mm) tend to have a relatively long working life, and they slide through descenders at a slower, more controllable rate. In comparison, thinner (e.g. 9 mm diameter) ropes are lighter for carrying to remote forest sites and are also easier to pull up into the canopy. The length of rope used depends on the height at which the researcher will be working. For maximum versatility, a 120-m-long rope will allow access to crowns of trees up to about 50 m in height.

Personal climbing equipment

The researcher is secured to the rope via personal equipment. The main item is the waist (sit) harness, of which there are several types. Ones made for rock climbing are designed to stretch and absorb the stress of a fall. Whereas caving harnesses

Table 1 Summary of the equipment needed for SRT, with suggested sources and costs

ITEM	NUMBER	MAKE[1]	CURRENT COST (US$)[2]
Rope (100 m) (10.6 mm)	1	Blue Water, PMI	205
Rope (10 m)	1	Blue Water, PMI	25
Sling (2–"24, 2–"48)	4	Blue Water	6–11
Harness	1	Petzl	40–80
Maillon (screwlink)	1	Petzl	17
Carabiner	10	Petzl, SMC	14
Cow-tails (10.5 mm)	3 m	Blue Water	8
Chest ascender	1	Petzl	40
Chest harness	1	Petzl	3–37
Handled ascender	1	Petzl, PMI	48
Foot loop	1	Petzl	2
Descending device	1	Petzl, SMC, PMI	40–70
Helmet	1	Petzl, HB, Blue Water	50–75
TOTAL			$500–630

[1] Manufacturers:
BlueWater: BlueWater, Ltd, 209 Lovvorn Road, Carrollton, Georgia 30117, USA
(www.spelean.com.au/BW/BWindex.html)
HB: HB Climbing Equipment, Bangor, Gwynedd LL57 4YH, UK (www.hb.wales.com/)
Petzl: Petzl Sa, Distribution Sport Zone Industrielle, 38920 Crolles, France (www.petzl.com/)
PMI: Pigeon Mountain Industries, Inc, P.O. Box 803, LaFayette, GA 30728-0803, USA (www.pmirope.com/)
SMC: Seattle Manufacturing Corp., 6930 Salashan Parkway, Ferndale, WA 98248, USA (www.smcgear.net/)
[2] Approximate: based on retail prices (2002)
We recommend equipment that is lightweight, durable and relatively inexpensive.

stretch less and allow for a more efficient progression up the rope. A harness allows the climber to be securely attached to other equipment (ascenders, descender, and safety lines). The waist harness can also be used to carry small items of useful equipment for later use. Descending and ascending equipment is attached to the waist harness by a screwlink (maillon), which rarely needs adjusting and is designed to be hard to open. The screwlink should never be opened during a climb.

During the climb, the researcher uses two types of ascender. These act like ratchets, allowing the rope to slide through when moved upward but clamping tight onto the rope when climber applies weight. This type of ascending is called 'frogging' (Smith & Padget 1996). There are different ways in which the two ascenders can be arranged. We recommend the arrangement of ascenders shown in Figure 1, which is favored by European climbers and is very efficient. In the SRT arrangement shown in Figure 1, the chest ascender is attached to the waist harness by a screwlink. A chest harness can be made with tubular nylon webbing, tied with a water-knot, crossed behind the climber's back, and attached to the chest ascender with a small screwlink. The chest harness simply holds the chest ascender in position and moves it up the rope when the climber stands. A foot ascender allows the climber to stand up, pulling the chest ascender up. Unlike the chest ascender, the foot ascender is moved up the rope manually during climbing (see below). A foot-loop is attached to the foot ascender by a screwlink.

An additional and essential part of the safety equipment is a pair of so-called cowtails. These are short lengths (e.g. 50 cm and 35 cm) of 11-mm diameter rope attached via the screwlink on to the waist harness. The free end of each cowtail has a loop (made from a figure-of-eight knot) which carries a carabiner. Carabiners are clips that are designed to be opened and closed during a climb to make temporary, secure attachments. The longer cowtail should be attached to the foot ascender (which is otherwise not secured to the climber) during a climb. This is a precaution against the remote possibility of failure of the chest ascender. Cowtails are also used as safety lines by allowing the climber to clip onto temporary anchor points (e.g. slings) in the canopy. This technique is essential when the climber moves off the main support rope in the canopy (see below). When not in use the cowtails can be clipped out of the way, to the side of the waist harness.

The researcher lowers himself/herself to the ground using a descender. A type of descender that works well is the auto-lock type. This allows the rate of descent to be easily controlled. The transfer of support between ascenders and the descender is critically important, and is discussed below.

Getting a rope into the canopy

(see chapter on line insertion on page 24)

Anchoring the rope

In SRT, the single most important step is in anchoring one end of the rope to at least one secure point (a belay). The anchor is the single, double or triple point at which the rope is tied to a tree or some other non-moveable object. The position of the anchor point depends on where the researcher first begins to use the rope. If SRT is used to climb from ground level, the anchorage point(s) will typically be the base of the tree being climbed. If SRT is being used to supplement other access methods, the anchorage point(s) may well be in the canopy.

To make an anchorage point, a loop (about 20 cm in diameter) is made of rope (usually but not necessarily at one end) by tying a knot. Probably the best knot for this is the figure-of-eight, because it is strong yet can easily be untied. Two carabiners are clipped into the loop. The rope can then be attached to a tree in two ways. One is to pass the rope around the tree and clip it onto itself; this method is particularly suitable for large-diameter trees. For smaller trees, we recommend using a sling. This is looped through itself to form a loop (a girth hitch), which the rope is clipped onto.

If ascent was by some other access method, it is possible to descend (by rappelling) to ground level on a rope. In such cases, there are two options for anchoring the rope. One way is to anchor the rope at ground level (as above) either before making the ascent or afterwards (by an assistant on the ground). The free end of the rope is carried or pulled up into the canopy and looped over a suitable branch.

The other method is to anchor the rope in the canopy. The best part of the tree to anchor to is the stem, particularly since branches are thinner in the upper part of the crown. Using this method, it may be necessary to take the rope down once the climber is on the ground. The rope can be retrieved with a remote belay release (Figure 2). This is simply a second rope or a cord that releases the rope and pulls it to the ground. The anchorage is temporary though safe because it can only be taken down when the climber's weight is off the rope (i.e. not pulling down at 'a'). To reduce the unlikely possibility of the upper carabiner (see 'c', in Figure 2) accidentally opening a second carabiner can be used, with the opening facing in the opposite direction. Warning – when a long length of heavy rope is released from the canopy it can fall to the ground beneath the tree with considerable force. It may be necessary to be positioned away from this area while operating the remote release.

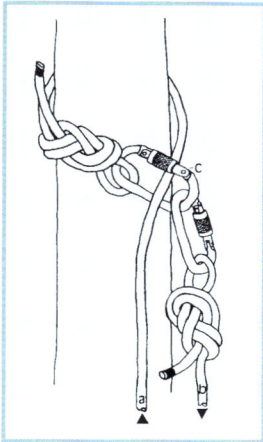

Figure 2. Detail of the remote belay-release.

This is used to retrieve a support rope in the canopy, by pulling from ground level. When the release rope is pulled at 'b', the support rope, 'a', is pulled through the upper carabiner 'c' and falls to the ground. When in use, carabiner 'c' bears the weight of the climber. Note the 'figure-of-eight' knots.

Ascending and descending the main rope

We do not provide detailed guidance here on climbing technique, because it can only be properly learnt by practice, under supervision by an experienced climber. Such practice should occur near ground level, where any mistakes may be embarrassing but not fatal. In any case, detailed instructions will not apply to all combinations of equipment and may therefore be confusing. However, it is possible to make a few general comments.

The most critical stages of rope climbing are the initial stages of ascent and descent. This is when support equipment is brought into use. Before ascending or descending a rope, all critical attachments should be double-checked, preferably (at least for ascents) by a colleague. The waist harness must be buckled correctly (manufacturers give advice on how to do this) and then never unbuckled until the climber is back on the ground.

Climbing cannot begin until the elastic slack is taken out of the rope. Even a static rope (see above) may initially have 1–2 m of slack, especially if the support branch is high in the canopy. Using the configuration shown in Figure 1, slack can be taken up by pulling rope through the chest ascender, using the foot loop to help the pulling. If there are any problems with any of the support points (e.g. waist harness, ascenders, overhead branch) they will probably become apparent during this stage. This is when any weaknesses in the system can still be corrected.

Ascent by SRT involves using the chest ascender and foot ascender alternately. Each ascender bears the climber's weight while the other ascender is being slid (manually or automatically) up the rope. The movement is similar to that of a jumping frog because the climber thrusts both legs down together to move upwards. The climber's legs should do most of the lifting work and the main thrust should be downward and backward. Tiredness in a climber's arms during a long climb is a sign of poor technique. The climber should be able to remove their hands from the climbing equipment at any stage in the climb and still feel secure.

During the climb, the chest harness merely pulls the chest ascender up when climbing and holds the climber in a comfortable semi-upright position. It is not part of the main support system.

A climber can descend at any point on the rope. The transfer from ascenders to the descender is technically challenging and must be rehearsed repeatedly near ground level before the climber commences the first climb. The main steps involve 'threading' the loose part of the rope (below the chest ascender) through the descender, which has previously (before starting the climb) been clipped securely to the screwlink on the climber's waist harness. All slack rope must be taken up, so that the climber can be sure that the descender is taking all his/her weight *before* descent is started. This is a challenging manoeuvre but is not hazardous because the climber is still attached by a safety line (a cowtail; see above) to the foot ascender. Standing on the foot ascender, it should be possible to release the chest ascender at this stage. When it is certain that the descender is taking the climber's weight as it should, the foot ascender can be unclipped and the descent can begin.

Rappelling down a long rope can be exhilarating, especially after a long session of sampling in the canopy. However, the climber should resist the temptation to descend too fast, in case of collision into vegetation and the possibility of losing control. The rope can be lightly guided through the descender, but hands should be kept clear to avoid trapped fingers in the mechanism, and the rope should not be gripped, to avoid burns.

Moving off and onto a rope in the canopy

Canopy researchers using SRT often need to move off the main support rope. There are two reasons for this. Firstly, the support branch that is used to first climb the tree may be low in the crown Secondly, sampling often needs to occur in the outer crown, where leaves, flowers, and associated fauna are mostly located (Nadkarni & Longino 1990, Lowman *et al* 1993, Moffett & Lowman 1995, Hadwen *et al* 1996).

The most critical stages of climbing within the crown are moving off and onto the main rope. This is, in a way, equivalent to the most hazardous part of flying a plane: take off and landing. Procedures must be followed carefully, since the climber shifts anchorage point from one part of the tree to another. During these transfers a safety line must secure the climber at all times.

Transferring off the main rope is easy if the rope is not on the lowest branch in the tree. This is because the climber can take his/her weight off the rope by standing on a branch below. The ascenders can then be easily removed (see Laman 1995). However, in many cases the main rope is supported by the lowest branch. In this situation, the climber needs to get up onto to that branch before moving higher into the canopy. To do this, it is possible to make a temporary foot support by securing a loop of rope or sling that hangs under the support branch. The climber can then step up on to the support branch. An easier technique is to use an atrier, which can either be bought or made. The atrier acts as a kind of ladder and allows the climber to step up to a higher position. These procedures are arduous but safe since the climber uses a safety line clipped to the tree at all times.

Techniques for moving around within the canopy depend on the architecture and size of the tree, and on which parts of the crown need to be reached. For all trees, the strength of branches decreases upwards and outwards, the climber should thus stay close to the bole. There are techniques for bringing the outer parts of the crown to the climber (see below), so that safety is not compromised.

Some trees have numerous stout branches arranged vertically within the crown. In such cases, the climber can ascend and descend the inner crown using the branches as a sort of ladder. However, the climber should clip on (using a cowtail) to a sling, to provide security in case a branch fails. The climber moves by clipping the other cowtail on to second sling placed further up or down the tree, using girth hitches around the bole or branches. Alternating between the two slings, the climber can ensure that he/she is always attached to at least one anchorage point.

Other trees have branches that are vertically widely spaced, which cannot be scaled easily or safely without a rope. In these cases we advocate using SRT for climbing within the crown. It is possible to move still higher in the crown by using the same rope repeatedly, provided that the climber uses a safety line to clip a sling on to the tree while the rope is being repositioned.

To ascend a large, open tree crown, it is necessary to place a short (e.g. 10-m) rope over a higher branch. Two such ropes (e.g. 9-mm diameter) can easily be carried in a rope bag. Getting the rope over the branch from immediately below is not always easy. However, it is possible to achieve this either with a throw bag (Whiteacre 1981) or by using an extendable pole pruner to push the rope over. If there is any doubt about the strength of thinner, higher branches, this method should not be attempted. Once the rope is passed over a branch, it can be anchored to the bole (see above) then climbed. A stopper knot *must* be tied near the end of this rope, to prevent the climber rappelling off the end of the rope when returning from the upper canopy. The knot must be large enough so that it will not pass through the descending device. There should be at least 60 cm of rope beyond the knot for added security. If necessary, a second rope can be used as a remote belay release (see Figure 2) once the climber is safely anchored near the main rope.

Some lateral movement is also possible within the crown, using a taut rope with a high support point to give maximum stability (Lilly 1994). The climber can walk outwards on sturdy branches. Lateral movement can also be achieved using running belays, in a method described by Whiteacre (1981).

Returning to the main rope is a critical stage, possibly made more difficult because the researcher may be tired after a long period of sampling in the upper canopy. Having left the main rope previously, it is *essential* that the climber knows which side of the rope to descend down. This must be on the opposite side from the anchored side, which can be checked by firmly pulling on the rope. At this stage, slack (generated when the climber moved off the rope) must be taken in. A safety line is used to attach the climber to the tree whilst transferring back to the main rope. The safety line should remain attached until the climber is confident that the descender is attached correctly.

A maritime line-firing gun adapted for use in the canopy. A small explosive charge fires the red ball and fishing line up over 80 meters. Photo: Andrew Mitchell.

Finally in this section, we want to re-emphasize the importance of safety when moving off or onto the main rope. The procedures above are described for information only. It is absolutely essential to practice these transfer techniques at ground level, with an instructor, before they are used in the canopy.

Extending the sampling range from SRT

There are methods by which a researcher can reach the outer canopy by SRT even if climbing within a tree's crown cannot achieve this. The suitability of method depends on whether sampling in the outer canopy is to be *in situ* or *ex situ*.

For *in situ* sampling, the researcher needs to make measurements in (relatively) undisturbed parts of the outer crown. An example of this is measuring stomatal conductance of leaves (e.g. Barker & Pérez-Salicrup 2000). One technique for reaching the outer canopy is to work from an adjacent tree. Another method is to work from a system of ropes attached to adjacent trees (Perry & Williams 1981). These approaches require the use of support trees (e.g. emergent trees) that are taller than the study tree. A possible disadvantage of using taller trees is that they are likely to shade the sampling area, at least for part of the day. For some (e.g. plant ecophysiological) studies this could be an unacceptable problem.

A very effective sort of *in situ* method is to bring an outer part of the crown to the climber by bending a branch. This might, for example, allow measurements of leaf gas exchange using an infrared gas analyzer that is securely mounted next to the researcher. If this procedure does not significantly alter the experimental conditions of the sample material, it may be an acceptable compromise.

For *ex situ* sampling, the challenge is simply to obtain material from the outer canopy, for measurement elsewhere. An example of this is collecting leaves for nutrient analysis (e.g. Barker & Becker 1995, Barker & Pérez-Salicrup 2000). Such material can be obtained remotely (and safely) from a rope using nets or pole-pruners (Mori 1995, Wieringa 1996). It is obviously important to keep sharp tools away from support ropes.

Limitations of canopy access by rope

We have outlined some of the advantages of using rope access for canopy research, but there are some disadvantages too. Most of the problems of rope access concern the effect of limited mobility on spatial or temporal aspects of sampling. The sampling zone from a rope is primarily restricted to a narrow vertical cylinder.

Lack of mobility of rope access has consequences for replication. Even when the outer canopy can be reached, obtaining sufficient replicates can be a problem. It is difficult to reach many independent samples (e.g. leaves or flowers) during one rope climb. One solution is to use trees as replicates, and to achieve adequate replication by multiple climbs. For time-sensitive sampling (e.g. studies of microclimate) this may introduce another variable, which is variation of environmental conditions between different sampling periods. However, this is a problem with other access techniques, and the fact that rope equipment is relatively inexpensive allows the possibility of multiple simultaneous sampling by more than one climber (Lowman & Bouricius 1995). For some studies, sampling equipment (e.g. mammal traps, microclimate sensors) can be raised and lowered using a pulley system installed by an initial rope climb (e.g. Taylor & Lowman 1996, Vieira 1998).

Disturbance through the use of ropes may be a problem in some studies and can affect areas of the canopy that are not being studied. For example, ropes can dislodge epiphytes which may affect other research, particularly in fragile forests. Disturbance by rope climbing (and those of other access methods) may also cause nest abandonment. However awareness of these possibilities allows creative solutions to some of the problems. For example rope climbing can be avoided at dawn or dusk, when birds are most active.

There are also practical (cf. experimental) difficulties in using a rope access system. Firstly, unlike some other canopy methods (e.g. tower, walkway, crane), rope access requires skill and fitness. Secondly, it is difficult to use heavy and/or delicate equipment while working from a rope in the canopy. Thirdly, it can be very uncomfortable to be suspended from a rope for relatively long (>1-hour) periods. A cheap, lightweight child's swing seat can make an excellent bosun's chair to aid mobility and comfort. These structures do not replace the safety systems of SRT equipment, which continue to provide the main support and safety.

Finally, we need to recognize the problems inherent with rope climbing that have no solutions. For example, ropes can only support a researcher from above, thus not allowing access to the top-most part of the forest, where the canopy-atmosphere interface occurs. Also, rope access often relies on the availability of large, healthy trees with convenient branches for support. Consequently, there may be problems in finding suitable support trees in stands that are dominated by mainly young (e.g. sapling- or pole-sized) or senescent (e.g. snag) trees. These limitations may prevent rope access in certain stands, including some secondary forests, where even mature trees may be relatively small.

Safety

Safety in rope climbing is limited by the most unpredictable part of the support system. If the climber is inexperienced, safety may be undermined by insecure anchorage (e.g. incorrectly tied knots) or incorrectly buckled waist harness. If the rope has been passed over branch whose strength is uncertain, that part of the system may fail. If the rope has an unknown history, it may break. The failure of any of these elements may have potentially fatal consequences, especially if it occurs during critical parts of the climb.

For some research projects, safety issues may extend beyond the immediate concerns of the individual climber. For examples, some research institutions may restrict certain canopy techniques, because of liability regulations. In some cases, due to legal obligations or simply a desire for good practice, projects may only use specialized climbers. Guidance on using SRT for canopy research has been published previously (Perry 1978, Whiteacre 1981, Mitchell 1982, Laman 1995).

Training

It is essential that all prospective canopy researchers are trained in the techniques of using ropes to access the canopy. We strongly advocate the use of SRT technical books such as *On Rope* (Smith & Paget 1996), *Vertical Caving* (Meredith & Martinez 1986). This information, coupled with professional training, will provide the necessary technical knowledge and problem solving ability to understand SRT. Finally, we strongly recommend the establishment of short in-forest training courses, promoting safe and effective canopy access and sampling by both new and

experienced researchers. Currently, there is no regular formal canopy access training program on the use of SRT. If canopy research is to grow, its development must be accompanied by still higher standards of safety and technique. There are few enough canopy researchers already. We look forward to being joined by others who understand the challenges and opportunities of using rope to reach the canopy.

Useful tips

- try to avoid disturbance to forest structure and function during access activities.
- become an experienced and safe climber who is aware of the limits of rope technique.
- ensure that sampling procedures are scientifically rigorous.
- be prepared to be inventive and creative with access and sampling techniques, provided that it is safe.
- allow enough time to prepare for the climb. Then double it.
- climb with a partner whenever possible and check each other before getting 'on rope'.
- make sure that equipment is not damaged and that you know its use history.

Never:

- put science before safety.
- assume. Recheck every component of the system if there is any doubt about its reliability.
- climb ropes when unwell or tired.

Contact details

Martin Barker & Noah Standridge, School of Forest Resources and Conservation, University of Florida, USA

Literature Cited

Barker MG (1997). An update on low-tech methods for forest canopy access and on sampling a forest canopy. *Selbyana* 18: 16–26

Barker MG & Becker P (1995). Sap flow and sap nutrient content of a tropical rain forest canopy species, *Dryobalanops aromatica*, in Brunei. *Selbyana* 16: 201–211

Barker MG & Pérez-Salicrup D (2000). Comparative water relations of lianas and mahogany (*Sweitenia macrophylla* King) canopy trees in sub-humid tropical forest in Bolivia. *Tree Physiol* 20: 1167–1174

Barker MG & Sutton SL (1997). Low-tech techniques for forest canopy access. *Biotropica* 29: 243–247

Barker MG & Pinard MA (2001). Forest canopy research: sampling problems, and some solutions. *Plant Ecology* 153: 23–38

Bongers F (2001). Methods to assess tropical forest canopy structure: an overview. *Plant Ecology* 153: 263–277

Cosson J.-F (1995). Captures of Myonycteris torquata (Chiroptera: Pteropodidae) in forest canopy in South Cameroon. *Biotropica* 27: 395–396

Dial R & Tobin SC (1994). Description of arborist methods for forest canopy access and movement. *Selbyana* 15: 24–37

Ellwood MDF & Foster WA (2001). Line insertion techniques for the study of high forest canopies. *Selbyana* 22: 97–102

Hadwen WL, Kitching RL & Olsen MF (1996). Folivory levels of seedlings and canopy trees in tropical and subtropical rainforests in Australia. *Selbyana* 19: 162–171

Hallé F (1990). A raft atop the rain forest. *Natl. Geogr.* 178: 129–138

Laman TG (1995). Safety recommendations for climbing rain forest trees with single rope technique. *Biotropica* 27: 406–409

Lilly S (1994). Tree Climbers' Guide. International Society of Arboriculture, Savoy, IL, USA

Lowman MD (1997). Herbivory in forests –from centimetres to megametres. pp 135–149. In: Watt AD, Stork NE & Hunter MD (eds), *Forests and insects*. Chapman and Hall, London

Lowman M & Bouricius B (1995). The construction of platforms and bridges for forest canopy access. Selbyana 16: 179–184

Lowman M & Moffett M (1993). The ecology of tropical rain forest canopies. *Trends Ecol. Evol.* 8: 104–107

Lowman M, Moffett M & Rinker HB (1993). A new technique for taxonomic and ecological sampling in rain forest canopies. *Selbyana* 14: 75–79

Meredith M & Martinez D (1986). *Vertical caving* (2nd edn.). Lyon Equipment, Dent, Sedbergh, Cumbria, UK

Mitchell AW (1982). *Reaching the rain forest roof. A handbook on techniques of access and study in the canopy*. Leeds Philosophical and Literary Society, Leeds, UK. pp 36

Moffett MW (1993). The tropical rain forest canopy: researching a new frontier. *Selbyana* 14: 3–4

Moffett MW & Lowman MD (1995). Canopy access techniques. In: Lowman MD & Nadkarni NM (eds.) *Forest canopies*. pp 3–26. Academic Press, San Diego

Mori SA (1995). Exploring for plant diversity in the canopy of a French Guianan forest. *Selbyana* 16: 94–98

Nadkarni NM (1988). Use of a portable platform for observations of tropical forest canopy animals. *Biotropica* 20: 350–351

Nadkarni NM & Longino JT (1990). Invertebrates in canopy and ground organic matter in a neotropical montane forest, Costa Rica. *Biotropica* 22: 286–289

Nadkarni NM & Parker GG (1994). A profile of forest canopy science and scientists –who we are, what we want to know, and obstacles we face: results of an international survey. *Selbyana* 15: 38–50

Perry DR (1978). A method of access into the crowns of emergent and canopy trees. *Biotropica* 10: 155–157

Perry DR & Williams J (1981). The tropical rain forest canopy: a method providing total access. *Biotropica* 13: 283–285

Smith B & Padget A (1996). *On Rope: North American Vertical Rope Techniques*. National Speleological Society, Huntsville, AL, USA

Taylor PH & Lowman MD (1996). Vertical stratification of the small mammal community in a northern hardwood forest. *Selbyana* 17: 15–21

Vieira EM (1998). A technique for trapping small mammals in the forest canopy. *Mammalia* 62: 306–310

Whiteacre DF (1981). Additional techniques and safety hints for climbing tall trees, and some equipment and information sources. *Biotropica* 13: 286–291

Wieringa JJ (1996). Tree-climbing on a free rope. pp 819–821. In: van der Maesen LJD, van der Burgt XM and van Medenbach de Rooy JM (eds). *The biodiversity of African plants*. Kluwer Academic Publishers, Dordrecht

Techniques for Inserting and Positioning Lines in Forest Canopies

By
Martin Ellwood
and
William Foster

MANY studies of the rainforest canopy can be conducted using ropes, even those that do not involve climbing. Alternative methods are often expensive, or arduous and slow. Studies that can be completed using ropes are usually inexpensive and versatile; they permit high levels of replication, and they can be easily repeated anywhere in the world. The value of skills developed while working with ropes will remain undiminished for many years. Whether for climbing or pulling up equipment, it is essential to be able to insert lines into the canopy. Lines have been used in a wide range of sampling studies including sticky traps (Compton *et al* 2000), fogging (Stork & Hammond 1997) and aerial netting (Munn 1991). Lines have also been used in the construction of walkways (Lowman & Bouricius 1995, Reynolds & Crossley 1995) and platforms (Nadkarni 1988, Workman 2000). However, forest canopies are structurally complex and unless lines are inserted with some precision, many areas will remain impossible to reach. The choice of technique will depend upon tree architecture and which part of the canopy needs to be studied. Here we will (1) briefly mention some line-insertion techniques and outline the problems associated with them, and (2) describe an efficient and versatile method for inserting and positioning lines that we have developed in Borneo over the past few years (Ellwood & Foster 2001).

Line-insertion techniques

Each kind of line insertion tool has its own strengths and weaknesses (see Table 1). In low canopy (below 20 m) it is possible to throw ropes over branches by hand. Specialist knots can be tied in the end of the rope, which unravel after the rope is thrown, allowing it to fall neatly over the nearest branch. Knots such as these are described in tree-climbing manuals (e.g. Jepson 1997). Alternatively, the rope can be attached to a small bag full of sand or equivalent known as a throwbag which when thrown pulls the rope with it (Dial & Tobin 1994). Throwbags have the advantage that they can be carried during climbing and used repeatedly to place ropes over branches in any direction. They are also inexpensive and safe to use, but they do have a limited range.

Hand catapults can also be used to insert lines as high as 30 m (Nadkarni 1988), whereas pole catapults are more powerful and can fire lines to heights of over 40 m (Munn 1991). Line-throwing guns are very powerful but they are also costly and potentially dangerous. Moreover, legal complications can make them difficult to import because they use explosive charges. Line-throwing guns are also non-adjustable and therefore not suitable for use in the lower canopy. Crossbows are highly accurate (see Table 1) and they can be adjusted, although they are also potentially dangerous. Transporting crossbows is easier than line-throwing guns because they do not use explosives and they can be dismantled and carried as component parts.

Table 1 The relative merits of various line insertion tools

TECHNIQUE	RANGE (M)	ACCURACY (M)	OPERATION	SAFETY	COST
Throwbag	15	± 5	easy	very safe	low
Pole catapault	40	± 10	moderate	unsafe	medium
Hand catapault	25	± 5	easy–moderate	safe	low
Line gun	80	± 2	difficult	hazardous	high
Crossbow	65	± 1	difficult	hazardous	medium–high

Techniques for inserting and positioning lines

Although working in forest canopies can be extremely challenging, it is the scientific requirements of an individual project that should govern where a study takes place, rather than the ease of access (Barker & Pinard 2001). For lines to be of use, they must be inserted with confidence and precision wherever needed. First, an optimum site within the canopy should be located. Second, a position that offers the best chance of successfully inserting a line into a predetermined area should be found. The point from which the line is eventually fired must afford a reasonably clear view of the branch, although most projectiles can be fired through light vegetation. A precise measure of the distance to the target site can be recorded instantly with a laser range finder. We use a Bushnell Yardage Pro™ 400, from Bushnell Sports Optics Worldwide, Kansas, U.S.A. This instrument records distances instantly, and many measurements can be made in a few seconds.

For work in high canopies (above 40 m), crossbows are almost certainly the best line insertion tools. Throwbags and catapults, although easy to transport and use, are not powerful enough to reach the base of the crown. Even if low branches are available, there is little to be gained from inserting lines halfway up a tree, unless that is sufficient for the study. It is possible to reposition lines from the ground (Figures 1 and 2), although more powerful insertion tools that are able to shoot further provide greater flexibility. In studies that involve climbing, firing lines into the highest branches makes the initial climb much easier. A relatively large, high-velocity (ca 76 m s^{-1}) crossbow is needed to fire an arrow with a line attached over branches above 40 m. The Panzer II crossbow, purchased from Barnett International, Wolverhampton, U.K., is sufficient for this purpose. Although detachable limbs make it possible to adjust the range of some crossbows, the process can be difficult and time-consuming under field conditions. Greater flexibility can be achieved by using arrows with a range of tip sizes (Table 2). The arrow shafts are 40 cm long, and weigh 16 g before the brass tips are attached. Brass tips can either be added permanently, or the shaft can be designed with a screw-on tip that can be changed in a matter of seconds, thus reducing the number of arrows required. The arrows are standard and can be purchased from Barnett International. The line is screwed into the back of the arrow (Figure 3), and the crossbow loaded. The reel can be held in the hand that supports the weapon, and this will allow the spool to unwind freely as the arrow is fired. Once loaded the crossbow should be fired without delay, otherwise a negligent discharge could cause serious injury. The crossbow and reel must be held in the firing position until the arrow has stopped moving otherwise the line will snap or tangle.

The height to which the arrow ascends depends upon the weight of its tip. Four sizes provide adequate flexibility (Table 2). Lighter arrows are required for high

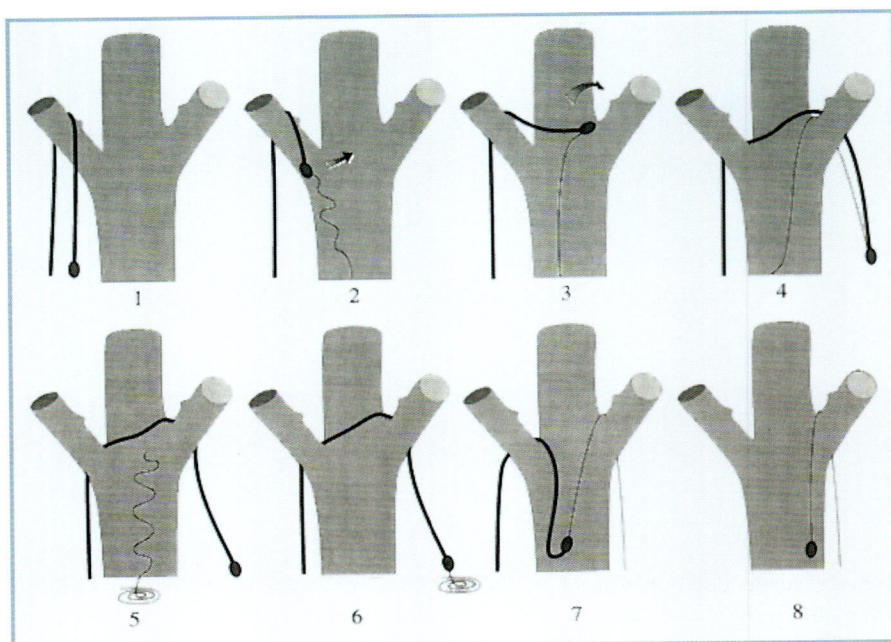

Figure 1 A ground based method for repositioning lines in trees.
A smooth weight attached to a thin line should be fastened to the thick line and pulled up to within a few metres of the left hand side (LHS) branch (1). Oscillations in the thin rope cause the weight to act as a pendulum and eventually to swing over the right hand side (RHS) branch (2,3). At this point (4) the weight is released and it falls, pulling both lines over the RHS branch. The thin line is pulled out of the tree (5) and re-tied to the weight (6), which is pulled back over the second branch (7). The weight, with both lines attached, falls down between the two branches, leaving the thick line over the LHS branch and the thin line over the RHS branch (7). The thick line can either be left in the tree or removed (8).

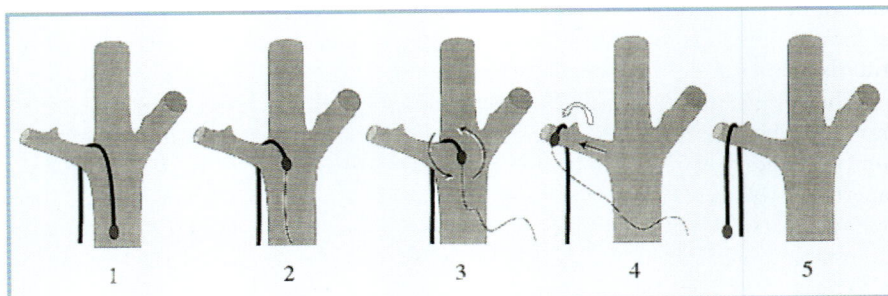

Figure 2 A ground based method for moving a line along a branch.
A weight is raised to the branch (1,2) as in Figure 1. But instead of oscillating the thin line, it is whirled around, causing the weight to move with a rapid circular motion (3). The momentum of the weight can be used to lift the line and place it further along the branch (4). The line can be lifted over the stumps and other obstructions, including branches, using this technique (5).

branches, but they must be heavy enough to pull the line over the branch and down the other side. The line is attached directly behind the shaft of the arrow (Figure 3). This prevents the arrow from straying during flight. If the line is not seriously tangled, the arrow should fall smoothly back to earth. If not, and if the

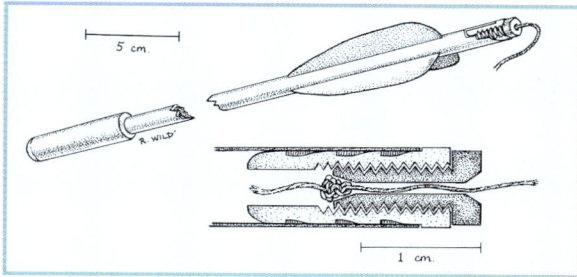

Figure 3 Crossbow arrow and line attachment system. Upper drawing shows modified brass arrow tip and line attachment system. Lower drawing is an enlarged longitudinal section showing the knot running centrally through the threaded brass rod, which is screwed into the end of the aluminium arrow shaft.

Table 2 Dimensions, weight and range of arrow tips

SIZE	TIP DIMENSIONS (MM)	TIP WEIGHT (G)	TOTAL WEIGHT (G)	RANGE (M)
1	60 x 15	74	90	30–40
2	50 x 15	54	70	40–50
3	40 x 15	44	60	50–60
4	40 x 12	24	40	60–65

arrow is suspended out of reach, it may be necessary to give the line a series of short, sharp flicks. In most cases, this will cause the line to slip and the arrow will drop. If the arrow has travelled an excessive distance past the target branch, it can be reeled back gently and allowed to drop vertically closer to the base of the tree.

Only thin lines can be successfully fired at high speeds, and they need to be smooth to reduce friction over the branches. Instead of nylon fishing line, braided line should be used because it has 'low memory' and is less likely to tangle. Using low-memory line is especially important when using high-velocity insertion tools. If the line twists, it will snap and the arrow will be very difficult to retrieve. The breaking strain should be high enough to be effective, but not so high that it cannot be broken if the arrow becomes entangled. For these purposes, 14 kg is sufficient. The line should be at least twice the height of the tree, in case the arrow travels further than the target branch.

The design specifications of the reel are important. Standard fishing reels will not work when used with high velocity projectiles. It is impossible for the line to unwind quickly enough and it will snap. It is essential to use reels with a shallow aperture, such as casting or spinning reels. The Daiwa Emblem spinning reel works well when loaded with Fox Pike System™ 14 kg line. Some fishing reels have an adjustable clutch that allows the unwinding speed to be controlled and this can also be used for controlling the flight of the arrow. Once over a suitable branch, the descent of the arrow can be controlled using the same principle. Having been lowered to eye level the arrow can be retrieved. If allowed to fall to the ground, it will become harder to locate amongst the vegetation. Once the arrow has been removed from the fishing line, a slightly thicker (ca 5 mm) line should be attached in its place. This intermediate line can then be pulled up into the canopy as the fishing line is wound back onto the reel. Finally, a heavy-duty (ca 10 mm) line should be pulled over the branch. The heavy-duty line can be used to pull climbing ropes or equipment into the canopy, and can remain there indefinitely. However, if lines are left in a tree for long periods, they may become choked with vines or lianas. They should therefore be checked every 3–6 months.

Lines can be moved between branches or along them. To do this from the ground, a smooth piece of wood weighing approximately 10 kg should be tied to a heavy-duty line and pulled up into the tree. This weight also should be attached to a slightly thinner line running back to the ground. Once in the tree, the weight can be moved back and forth by swinging the thinner line. By appropriate control of the movement of the weight, the line can be moved either to a new branch (Figure 1) or to a new position on the same branch (Figure 2). These techniques are easier to perform if the branches are clearly visible. The alternative is to fire another line although, if the branches are obscured, this option may be less desirable than persevering with the original line.

Contact details

Martin D.F. Ellwood, The Insect Room, Department of Zoology, University of Cambridge, Downing Street, Cambridge CB2 3EJ, UK. Tel: 01223 331768 (Direct Line). International: +44 1223 331768. Fax: 01223 336676. International: +44 1223 336676. email: mdfe2@cam.ac.uk

Literature cited

Barker MG & Pinard MA (2001). Forest canopy research: sampling problems, and some solutions. *Plant Ecology* 153: 23–28

Compton SG, Ellwood MDF, Davis AJ & Welch K (2000). The flight heights of chalcid wasps (Hymenoptera: Chalcidoidea) in a lowland Bornean rain forest: fig wasps are the high fliers. *Biotropica* 32: 515–522

Dial R & Tobin SC (1994). Description of arborist methods for forest canopy access and movement. *Selbyana* 15: 24–37

Ellwood MDF & Foster WA (2001). Line insertion techniques for the study of high forest canopies. *Selbyana* 22: 97–102

Jepson J (1997). *The Tree Climber's Companion*. Beaver Tree Publishing, Minnesota

Lowman MD & Bouricius B (1995). The construction of platforms and bridges for forest canopy access. *Selbyana* 16: 179–184

Munn CA (1991). Tropical canopy netting and shooting lines over tall trees. *Journal of Field Ornithology* 62: 454–463

Nadkarni NM (1988). Use of a portable platform for observations of tropical forest canopy animals. *Biotropica* 20: 350–351

Reynolds BC & Crossley DA (1995). Use of a canopy walkway for collecting arthropods and assessing leaf area removed. *Selbyana* 16: 21–23

Stork NE & Hammond PM (1997). Sampling arthropods from tree-crowns by fogging with knockdown insecticides: lessons from studies of oak tree beetle assemblages in Richmond Park (UK). In *Canopy Arthropods* (Eds, Stork NE, Adis J & Didham RK) Chapman and Hall, London, pp. 3–26

Workman M (2000). Sabah 2000: Exercise Pelopor. *Finn. Royal Geographical Society*, Report Number 3942, London

The Canopy Platform:
An Economical but Limited Access Technique

By
Bart Bouricius,
Austin E. Stokes
and
Brian. B. Schultz

THE canopy platform as a tool for scientific investigation has been around for at least 70 years (Mitchell 1986). This access technique offers improvements over the single rope technique because it allows several collaborators to work in close proximity to each other while carrying out research in the canopy. Although platforms may be used as a component of canopy walkway systems (where they are combined with connecting bridges), our remarks are mainly restricted to free standing platforms.

Platforms do have some advantages over canopy walkway systems, in terms of flexibility and cost. A walkway system cannot be easily moved, and is harder to justify financially, for short-term use (a number of months or weeks). However this is not the case for stand-alone platforms, which can easily be designed as temporary or permanent structures. The disadvantage of platforms is simply that they do not afford the degree of lateral movement (e.g. transects) that walkway systems with bridges permit.

Methods of site selection
Site selection involves:
1. Choosing a forest site appropriate to the research purpose, in which a sturdy/healthy tree of adequate size and species is selected.
2. Avoidance of proximity to large cleared areas, which may increase the chance of damaging high winds and may also decrease tree stability. If the research objectives require the use of such environments, guy cables can be used to stabilize the tree in which the platform is to be built.
3. Placement of the platform such that both structural and research criteria are met. An individual platform can be constructed in various ways such that it is in the high canopy branches, or on the bole of the tree. When an entire walkway system is being built, it is substantially more costly to place platforms in difficult locations on the tree. "Crows nests," which consist of wooden poles or aluminum ladders with seats and are stabilized with guy cables, may allow one to reach the uppermost branches or work above the canopy. For canopy net systems, the platform tree must also be chosen with an eye for potential net lanes, minimal disruption to vegetation, and anchor trees for attachment of net cables and control cords.

Permanent platform construction technique
When considering a platform, it is important to keep in mind that each tree with its ensuing construction challenges is unique. There are also always several possible sets of research objectives for which any particular platform may be built. With this in mind, it is impossible to dictate uniform, rigid standards for all construction sites. The only thing that can really be said is that any platform is structurally sound and has as little impact on the tree and surrounding environment as possible.

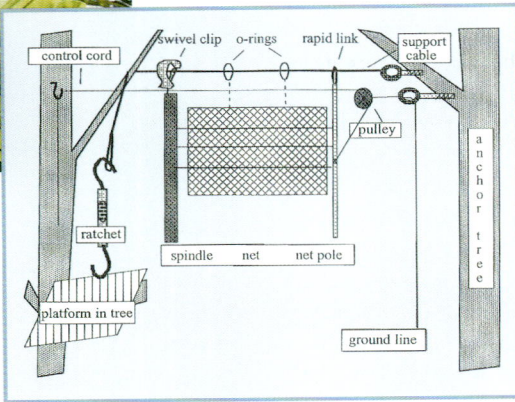

Figure 1 Net deployment from canopy platform.

For typical research, a 'permanent' platform is constructed above the point at which the crown limbs have branched out. This is done both to obtain maximum height for the platform, and to avoid placing constricting fabric webbing around or bolts in the main trunk of the tree.

In a typical scenario, two stainless steel or galvanized steel aircraft cables are placed horizontally in the lower crown. Three to five joists are then placed on and clamped to these cables. The decking is then installed on top of the joists, and the corner posts are securely fastened to the deck with necessary bracing. Additional posts for added safety may be placed on either side of the proposed entranceways to the platform. Holes are drilled in the posts to allow for 3 or 4 horizontal levels of 1/2" polyester rope to encircle the platform where there are no entranceways. Polyester is used because it has minimal stretch, and lasts longer than nylon in sunlight, while providing comparable strength. At this stage, holes are drilled in the deck and vertical ropes are spliced to the horizontal ropes every 12", thus providing a sturdy safety net fence (cf. Lowman & Bouricius 1995) for researchers. For extra safety, back-up cable loops are connected to the ends of each cable supporting the platform, and to the ends of the overhead cable to which a safety lanyard may be attached from the researchers harnesses.

Access systems for getting to the platforms can consist of ladders attached to the tree and/or a block and tackle system. All access systems should include a redundant safety rope or cable to which the user is attached while ascending. Ascending researchers may enter the platform through a hole in the deck, or over the side, depending on the structure of the tree, and the preferences of the researchers.

Research projects conducted

Research we have conducted at Hampshire College involves the netting and capture of nearctic migrants (Stokes & Schultz 1995, Stokes *et al* 2000). This research which is described below, also serves as a useful illustration of some of the pros and cons of using platforms. We used three platforms, one stand-alone and two connected by a walkway bridge. From each of the three platforms two mist nets were deployed at an average height of 21 meters above the forest floor. Our intention was to use nets in the canopy, in the same way as they have been used at ground level for many years. Platforms provided a safe, stable place for workers to deploy the nets (Figure 1). There is also the added advantage that only branches intruding on the net lane have to be trimmed, instead of trimming all branches from ground level up to net height as would be required when installing the nets from the ground. However the downside is that if movable platforms are used, a new net lane at the height of the nets must be cleared at each location.

Workers stationed in the canopy for long periods are able to remain comfortable and thus relatively still which generally means they are inconspicuous. This enables mist nets to concurrently sample two forest strata with a minimum of disruption.

During the spring and autumn migration seasons the bird species caught in canopy nets (which had more birds overall in the autumn), are generally canopy foragers (e.g. Red-eyed Vireo *Vireo olivaceus*, Scarlet Tanager *Piranga olivacea*, and migrant warblers). Whilst the bird species caught in understorey nets (which had more birds overall in the spring) were generally ground foragers (e.g. Gray Catbird *Dumetella carolinensis*, Ovenbird *Seiurus aurocapillus* and Wood Thrush *Hylocichla mustelina*. Of 40 species netted, 17 were captured in canopy nets and were not represented in conventional capture results from the understorey directly below these nets (Stokes, unpubl. data). Using this method, it could be possible to sample multiple strata using platforms at differing heights in the same (albeit very tall) tree.

One advantage of a single platform in a tree is that it can become an integral part of the landscape and "blend in" thus having little noticeable effect on study subjects. As opposed to other man-made structures, such as towers and cranes, which potentially introduce a high degree of artificiality to a site. On the other hand, at our temperate, deciduous forest site, nets and platforms were highly visible (and potentially avoided by birds) before leaf emergence in spring; this issue may be less of a problem in other situations, such as tropical or evergreen forests.

Although a single platform is sufficient to deploy nets, the walkway bridge was very useful in enabling one researcher to service a total of four nets without descending to the ground. In a complex walkway system the number of potential nets could be much greater.

Another component of the same research involved hanging winter bird feeders in horizontal layers in the canopy. The bridge itself, proved more useful than the platforms for attaching the bird feeders at high and low points to enable easy viewing from the ground (birds prefer higher feeders). The platforms attached to the bridge also made it possible to observe the birds from above and below the feeders (to show that our presence on the ground was not driving them higher).

Platforms both permanent and mobile, can potentially allow standard ornithological point counting methods to be used in the canopy, with replication occurring simultaneously at ground level. Thus allowing comparisons and tests against the accuracy of traditional ground-based sampling methods and observer bias.

Furthermore platforms offer 'outdoor labs' for students in science classes, enabling basic natural history observations of birds and mammals, and the placing of insect and mammal traps (not feasible with a bridge unless it is very solid and wide). The platforms and bridges are also invaluable as a tool for public education, making it possible for a group of adults or children to spend time in the canopy with maximum comfort and safety.

Canopy Construction Associates (www.canopyaccess.com), has been involved in constructing two stand-alone platforms, and several canopy walkway systems (e.g. Lowman & Bouricius 1995, Stokes & Schultz 1995), which involved building some 37 platforms in all. We have found that, for short term platforms where the tension on attachment points is low, webbing slings and 1/2" polyester ropes are useful for attaching platforms to trees. While for long term platforms we prefer to use the traditional arborist technique of bolting cables to trees, which avoids the possible girdling effects of encircling a tree under high tension. The precise method of attachment varies and is dependent on tree species, and the degree of tension that is created at the attachment point.

Costs

The cost of an individual platform may vary, depending on the dimensions of the platform, the location of the site, and the logistical constraints involved in moving materials and personnel to and across national borders. We have provided a description of the materials and costs necessary for a typical permanent platform. It would be pointless to get into more construction minutiae than this.

Table 1 Labor and materials for one 8' x 8' permanent platform

DESCRIPTION	SIZE	AMOUNT	COST US$
Polyester rope for safety barrier railings	1/2" diameter	300'	188
Pressure treated decking boards	2" x 8" x 8'	14	90
Posts for rope safety barrier fence	4" x 4" x 10'	3	25
Joists for deck support	4" x 6" x 10"	5	75
Braces for post stabilising	2" x 4" x 10"	7	25
Stainless steel cable for support and safety lines	3/8" diameter	100'	125
Copper swaging sleeves for 3/8" cable		14	42
Heavy drop forged cable clamps for 3/8" cable		10	15
Drop forged galvanized 18" eye bolts	3/4" diameter	4	80
Drop forged galvanized 15" eye bolts	3/4" diameter	4	60
Drop forged galvanized 14" eye bolts	5/8" diameter	4	28
Spiral galvanized nails boxes of 200	3.5"	2	12
Galvanized 3/8" x 5.5" hex lag screws		100	14
Carriage bolts 1/2" x 6" galvanized		50	15
Galvanized 3/8" x 1.5" hex lag screws		20	3
Galvanized bar washers		10	2
Labour (2 people x 3 days x $250/day)			1500
TOTAL			2303

Note: these figures only include labor and material costs, excluding other costs such as transportation and shipping.

Values are general estimates for a typical platform, not including transportation and shipping costs. Dimensions are in feet and inches (this is how materials are sold in the US) with costs rounded off to the nearest dollar.

Contact details

Bart Bouricius, Research Associate at Hampshire College Amherst MA, and Selby Botanical Gardens Sarasota FL. Founding member of Canopy Construction Associates: www,canopyaccess.com, 32 Mountain View Circle, Amherst MA 01002

Dr. Brian Schultz is an associate Professor at Hampshire College who teaches Ecology and Statistics

Austin Stokes is a consulting ornithologist from Pitsfield MA.

Literature cited

Lowman M & Bouricius B (1995). The Construction of Platforms and Bridges for Forest Canopy Access. *Selbyana* 16:179–184

Mitchell AW (1986). *The Enchanted Canopy*. Macmillan, New York.

Stokes AE & Schultz BB (1995). Mist Netting Birds from Canopy Platforms. *Selbyana* 16:144–146

Stokes AE, Schultz BB, DeGraaf RM & Griffin CR (2000). Setting Mist Nets from Platforms in the Forest Canopy. *J. Field Ornithology* 71:57–65

Sectional Aluminium Towers used in Forest Hydrological and Gas Exchange Process Studies

By
John Roberts

THE use of towers in various hydrometeorological studies commenced at CEH-Wallingford (formerly the Institute of Hydrology) in 1968. When the Institute was established its remit was to investigate the impact of conifer afforestation on water resources in the uplands of the UK. In parallel to catchment studies in mid-Wales, the Institute established a study in a mixed pine forest at Thetford Chase, East Anglia. The purpose of this study was to study evaporation processes in detail.

Several of the methods of studying evaporation and its controlling processes requires measurements of the radiation balance of the forest and of gradients of temperature, humidity and windspeed. At that time the instrumentation available meant that thermometers and anemometers needed to be mounted in a profile extending from a point within the trunk space and forest canopy well above the forest out of the air space influenced by the forest. At Thetford Forest which was composed of trees around 20m tall this meant forest towers which were usually about 33m high.

Clearly a vital situation of any tower is that it should remain vertical, stable and rigid until it is dismantled and removed. Towers are expensive if they fall over, the damage may not be repairable and if the tower is shaky, personnel are disinclined to carry out even routine tasks. If the towers are unstable and not upright some measurements, which require that instruments are horizontal and vertical, might be compromised.

The stability of the towers comes only partly from the method of construction with strong sections and bracing poles. It has been the practice from Wallingford to provide a stable base and guy wires for the towers.

Given that towers have to be erected to substantial heights and there has been a frequent research requirement, the erection procedure developed by staff from Wallingford has needed to be methodical but above all safe. Of course, the tower needs to be vertical and this is ensured by careful placement and levelling of the tower base and the first section.

Figure 1 illustrates one of the many and temporary uses the tower sections might be configured for.

Safety in use

Clearly a fall from any significant height can cause injury or death, so care needs to be made when climbing and using tall towers in forests. Some towers will have sufficient numbers of supporting tubes that a fall out of a tower is unlikely but extended falls can be avoided by using a central restraining rope and short rope waist harness.

There are more severe risks to slipping on towers when there is ice or snow on the platforms or when the wooden platforms are covered with algae and are wet.

Standing well above the forest canopy lightning strikes to the tower are quite likely so it is very dangerous to remain on a tower during storms when lightning is

Figure 1. A tower built in Niger, West Africa to measure fluxes from savannah vegetation.
(1) An eddy correlation device to measure evaporation fluxes. (2) Automatic weather station (3) Solar panels (4) Radiometers measuring crop surface reflection well away from the tower to avoid interference.
(5) One of 2 ends of the tower N section (6) Platform (7) Various connecting tubes. Photo: John Roberts.

occurring. It is important then not to begin work on a tower if storms are likely and to postpone on-tower activities that are in progress and descend to safety promptly but taking all the usual care used when descending the tower normally.

Contact details

Dr John Roberts, CEH-Wallingford, Wallingford, OX10 8BB, UK

Canopy Walkways – Highways in the Sky

By
Margaret Lowman,
Mark Hunter,
Bruce Rinker,
Tim Schowalter,
and
Steve Conte

RESEARCH in forest canopies has been limited by logistic constraints of access (reviewed in Mitchell 1982, Moffett & Lowman 1995). Over the past decade, several inexpensive techniques have been developed, but they are usually restricted to solo efforts. These include single rope techniques (looked at earlier); ladders (Selman & Lowman 1983, Gunatilleke *et al* 1994); and towers (Odum & Ruiz-Reyes 1970, Zotz 1994). Devices that facilitate research by a group of scientists simultaneously have also been developed, but are usually considerably more expensive (e.g. the raft and dirigible (Hallé & Pascal 1991)); construction cranes (Parker *et al* 1992)). In essence, there appears to be a distinct correlation between expense of access method and number of scientists that can safely utilize a common device (see Table 1 in Moffett & Lowman 1995).

Walkways offer an alternative means of studying forest canopies in a more comfortable, permanent fashion, thereby facilitating long-term and collaborative studies that are not feasible with ropes, or in cases where rafts and cranes are not affordable. With the modular system of design described below, it is possible to construct systems that allow scientists to replicate both within and between tree crowns, and to conduct repeated measurements over time and space. These modular systems, consisting of interconnected bridges and platforms (Figures 1 and 2), are of moderate cost and provide very easy access to users over a relatively long lifespan.

In 1983, one of the world's first canopy walkways was designed on the back of a cocktail napkin in the tropical rain forests of Queensland, Australia (Lowman & O'Reilly, pers. comm.). This was developed as an improvement to our original canopy access technique of solo climbing, which presented a challenge for group research efforts such as biodiversity sampling or Earthwatch supported expeditions using volunteer field assistants. After the walkway in Queensland was successfully funded and built by the Green Mountain Natural History Association, another walkway design was independently constructed in Malaysia (Ilar Muul, personal communication). Other walkways had been built for temporary or restricted research expeditions, but these were the first two permanent walkways designed for multiple use, incorporating ecotourism and research.

In 1992, one of us (ML) set up a non-profit company, called Canopy Construction Associates, with an arborist partner (Bart Bouricius). From this original partnership, the first canopy walkway in North America was constructed in the research forest of Williams College, Massachusetts. This walkway was a prototype, and each phase of the design was carefully costed. The walkway was also created in a modular fashion, with both the bridges and the platforms serving as units that can be replicated to create walkways of all shapes, lengths, and sizes. Since that first construction in a temperate forest, many scientists and researchers have consulted with Canopy Construction Associates for advice, planning and construction of canopy walkways. The process of canopy walkway design and construction are outlined here, with two sample budgets presented for different sized structures.

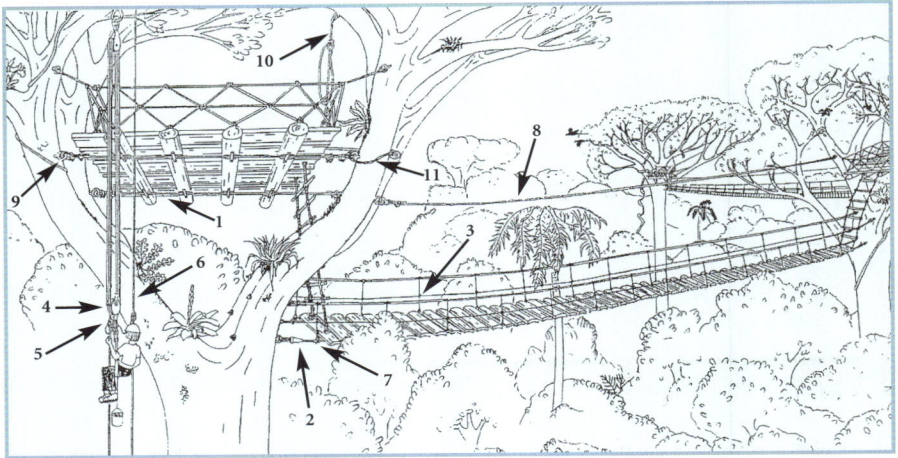

Figure 1 Illustration of the modular system of construction of bridges and platforms. Components shown include:

1. 3/8" stainless steel wire rope (12,000 lb. tensile strength);
2. Strand vice – a hardware item that allows precise measuring and tensioning of a cable while it is being installed;
3. Stainless steel net clamp – clamps two cables together at right angles;
4. Block and tackle – over 10,000 lb. tensile strength;
5. Ascender – device to prevent unintentional slide down rope;
6. Redundant safety rope – makes users feel safer during ascent;
7. Bridge clamp – clamps onto bridge support cable providing a connection to the vertical side cables and prevents unnecessary flexing of cables;
8. Cable for attaching safety lanyard to when walking on bridge;
9. 5/8" x 18" drop forged galvanized steel eye bolt (17,500 lb. tensile strength);
10. Adjustable rope safety lanyard;
11. Redundant cable provides extra security at all major connections.

Methods of Site Selection and Construction

Site selection must integrate both engineering and biological considerations. Engineering constraints include:

1. selection of a forest site of mature, healthy canopy trees within close proximity (walkways and platforms are not safe if built in trees that are small or show signs of crown dieback or trunk rot)
2. use of canopy trees with upper branch systems that are conducive to support of platforms
3. selection of a stand of trees with potential for expansion of modules (the minimum operational design consists of one bridge and one platform)
4. avoidance of close proximity to edges and treefalls, since these aberrations in the canopy create wind patterns that may lead to damage of the trees in the vicinity of the walkway

Biological considerations are equally important, when research is the major function of the structure. Biological factors include:

1. selection of a stand of trees that is representative of the species composition and diversity of the forest type

Figure 2
The first canopy walkway
constructed in North America,
at Williams College,
Massachusetts.
Photo: Meg Lowman.

2. placement of bridges and platforms to enable maximum access to foliage and crown space, but with minimal disturbance to the crown architecture
3. physical dimensions of the structure that are conducive for the intended research
4. rigorous standards of construction that minimize impact on the ground and the understory, as well as on the canopy

The minimum aerial construction module consists of one platform or one bridge. We have found that two platforms with an aerial bridge connecting them maximizes research opportunities for the cost. A slightly larger system will enable researchers to replicate both within and between tree crowns, which improves the rigor of ecological sampling. The bridges are strung between trees, with a maximum expanse of approximately 30 m. The hanging bridges consist of grooved aluminum or treated wooden treads attached to 3/8" galvanized steel cable of the type used in aircraft (14,400 lb. tensile strength). Hand rails are made with 3/8" GAC (galvanized aircraft cable) webbing between the rails and the ties are strung with 3/16" GAC with a 4,200 lb. tensile strength. The platforms are constructed of aluminum beams or pressure-treated wood suspended on the same 3/8" cable (referred to as multi-strand cable) used in the bridge construction. The platforms have 1/2" polyester combination rope webbing (6,000 lb. breaking strength), including hand rails. The webbing is strung between the platform floor decking and the rails.

This method of suspension construction has been chosen to avoid the possibility of structural members rubbing against the tree limbs when the trees move in the wind. This protects both the wooden structure and the tree from damage. The cable strength provides an extra measure of safety over other construction methods that might be considered. We have constructed several walkways successfully in different forest types using this precaution, because it minimizes impact upon the foliage, boles and tree architecture.

In temperate deciduous forests, we have constructed four systems: oak-maple-beech forest at Williamstown, Massachusetts (Figure 2); oak-maple forest at Millbrook School, Millbrook, New York; beech-hickory forest at Hampshire College, Amherst, Massachusetts; and three replicate sites of temperate deciduous forest at Coweeta Hydrological Reserve, North Carolina.

In tropical environments, we have constructed several systems, such as a site at Blue Creek, Belize which was filmed extensively as part of the Jason Project for Science Education, during 1994; it is now used both for ecotourism and research.

Considerations of Cost

We have costed our designs by modules, with the idea that different budgets and varying architectural features in a stand will determine the numbers of bridges and platforms to be utilized in a site. The materials have been priced separately from labor, because the latter will vary with location. The cost of materials may also vary if extensive shipping to remote sites is required.

The total costs of construction of four platforms and three bridges, plus three ladder systems in the subtropical rain forest of Belize was approximately $32,000 in 1994. Because this walkway system was being used for a major film production, several special features were included: rubber grommets between the rungs of the bridges, to minimize clanking of the aluminum; a Bosun's chair to demonstrate this special mode of canopy access; and an Eagle's nest above the canopy for observations of flowering events throughout the entire valley. The platforms were situated at a range of heights between 27 m and 42 m, and the bridges ranged in length from 12–25 m. In total, over 10 canopy trees and over 25 understory trees were accessible from this structure.

In contrast, a more modest access system at Millbrook, New York was completed for $20,000 in 1995. The platforms were situated at 10 m to 25 m height, and three bridges spanned 56 m in total. The costs for this system are itemized in the appendix.

Discussion

Although the rigors of winter are harsh for temperate sites, the problems of humidity and moisture in the tropical sites are probably more challenging. The use of stainless steel cable and rot-resistant wood is advocated for major support cables in all situations for reasons of safety and longevity.

There are two potential drawbacks to the use of bridges and platforms for canopy research. First, the structures are permanent and it is not feasible to shift them to new locations for purposes of comparing different sites. Second, there is a possibility that organisms in the canopy will utilize the bridges for enhanced mobility; however, we have not (in our five years of experience) observed any such behavior to date. (Lianas probably provide similar mobility.)

We are happy to discuss designs and give advice on any prospective canopy access project. We are also coordinating the collective maintenance of these structures, and a collaborative network of data collection that is developing at each site. The next decade should bring about the advent of a productive network of walkways, yielding comparative results about canopy processes.

Contact details

Dr Margaret Lowman, Marie Selby Botanical Gardens, 811 South Palm Avenue, Sarasota FL 34236 USA (mlowman@selby.org for correspondence)

Literature cited

Gunatilleke IAUN, Gunatilleke CVS & Dayanandan S (1995). Reproductive biology of Sri Lankan Dipterocarps. *Selbyana* 15(2): A9.

Hallé F & Pascal O (1991). *Biologie D'Une Canopée de Forêt Équatoriale – II.* Rapport de Mission: Radeau Des Cimes. Réserve de Campo, Cameroun. Avec Le Parrainage Du Ministére de la Recherche et de la Technologie.

Lowman MD (1984). An assessment of techniques for measuring herbivory: is rainforest defoliation more intense than we thought? *Biotropica* 16(4): 264–268.

Lowman MD (1998). *Life in the treetops: adventures of a woman biologist.* 224 pages. Yale University Press, New Haven CT.

Mitchell AW (1982). *The Enchanted Canopy: secrets from the rainforest roof.* William Collins Ltd, Glasgow, UK.

Moffett M & Lowman MD (1995). Methods of access into forest canopies. Pp. 1–24 in M.D. Lowman MD and Nadkarni N, eds., *Forest Canopies.* Academic Press, San Diego.

Nadkarni NM (1984). Epiphyte biomass and nutrient capital of a neotropical elfin forest. *Biotropica* 16(4): 249–256.

Odum HT & Ruiz-Reyes J (1970). Holes in leaves and the grazing control mechanism In: Odum HT and Pigeon RF eds., *A Tropical Rain Forest.* U.S. Atomic Energy Commission Rio Piedras.pp.I–69 to I–80.

Parker GG, Smith AP & Hogan KP (1992). Access to the upper forest canopy with a large tower crane. *BioScience* 42: 664–670.

Perry DR (1978). A method of access into the crowns of emergent and canopy trees. *Biotropica* 10(2): 155–157.

Selman BJ & Lowman MD (1983). The biology and herbivory rates of *Novacastria nothofagi* Selman (Coleoptera: Chrysomelidae), a new genus and species on *Nothofagus moorei* in Australian temperate rain forests. *Aust. J. Zool.* 31: 179–91.

Zotz G (1994). Ecology and demography of vascular epiphytes. *Selbyana* 15(2): A25.

APPENDIX 1. Budget for Canopy Access Structure at Millbrook School, Millbrook, New York (1995 prices)

This appendix costs out materials for 3 (8' x 8') platforms with rope hand rails and retaining rope-net, aluminum ladder access, and 168 feet of hanging cable supported bridge in 3 spans. (This original design was altered slightly to four platforms after arborists had examined the structural features of the upper canopy at the site; but the budget and equipment list remained the same.)

BRIDGES: This part of the walkway system consisted of 7 cables running between two trees. The two lowest cables held up the foot treads which were fabricated from pressure-treated lumber. (Other naturally resistant wood or aluminum are also sound materials.) The top cable, which is located about 7' above the bridge treads, is a 3/8" safety cable to which walkway users are tethered while they remain on the bridges. Two 3/8" cables were located 4' above the foot tread support cables and served as hand rails. Two more smaller cables were located half way between the foot cables and the hand cables; together, with vertical cables connecting the hand and foot cables, they provided steel nets on the sides of the bridge. Four additional cables, which were not part of the bridge structure, were used as guy wires to counterbalance the weight of the bridges on the trees.

PLATFORMS: The platforms consisted of 4" x 6" pressure-treated southern yellow pine joists with 2" x 6" pressure-treated decking all supported by two 14,200 lb tensile strength cables (and in some cases, four cables). The platforms had polyester rope retaining netting surrounding them and a security cable above which users can attach.

Table 1 Budget for canopy access structure at Millbrook School

MATERIALS

PIECES	DESCRIPTION	PRICE US$
34	3/4" x 18" DFG eye bolts	448
10	3/4" x 14" DFG eye bolts	125
4	3/4" x 24" DFG eye bolts	160
30	5/8" x 15" DFG eye bolts	522
18	5/8" x 18" DFG eye bolts	88
2	5/8" x 24" DFG eye bolts	61
12	5/8" x 24" DFG double arming bolts	72
86	3/8" x 3" GV U-bolts w/4 nuts/cross plates	172
42	1/2" x 3 1/4" DFG eye lags	180
16	5/8" x 6 3/4" DFG thimble eye lags	127
3	Spreader bars 18" x 4" x 1/2"	88
6	Pear-shaped sling links 1/2" diameter	44
108	Heavy galvanized thimbles for 3/8" cables	97
200	Heavy galvanized thimbles for 3/16" cable	60
24	Square DFG washers for 3/4" bolts	19
24	Round DFG washers for 3/4" bolts	6
30	Round DFG washers for 5/8" bolts	7
3000	Feet of 7 x 19 3/8" GV steel aircraft cable	1,740
600	Feet of 7 x 19 3/16" GV steel aircraft cable	150
85	Net clamps	170
85	Aluminium dead end clamps	850
170	Aluminium oval swedging sleeves	31
36	Feet of aluminium spacer tubing	90
270	DFG heavy cable clamps for 3/8" cable	648
30	DFG heavy cable clamps for 3/16" cable	54
108	Galvanized serving sleeves for 3/8" cable	125
30	DFG 1/2" diameter 6" staples	180
3	Type II 16' aluminum extension ladders	258
1	Type II 20' aluminum extension ladder	100
200	Feet of 1/16" galvanized seizing wire	15
1200	Feet 3 strand 1/2" combo polyester laid rope	1,000
100	Feet kernmantel polyester braided rope	120
3	5 lb. boxes galvanized twist nails	21
3	Small boxes galvanized long fence staples	9
15	4" x 6" x 8' pressure-treated beam joists	200
45	2" x 6" x 8' pressure-treated decking	270
70	2" x 4" x 12' for pressure-treated treads	455

TOTAL COST FOR MATERIALS   9,100

Shipping	*400*
Air travel, two persons	*600*
Other travel	*100*
Time spent planning and preparation	*800*
Cost of labour	*8,400*
TOTAL PRICE OF MAIN WALKWAY SYSTEM	19,440

Table 2 Additional costs that can be incurred for proper use of a walkway system

MATERIALS

PIECES	DESCRIPTION	PRICE US$
4	*ascenders for use as safety devices whilst climbing*	*160*
4	*blue water climbing helmets*	*176*
4	*fudge harnesses*	*96*
4	*double safety lanyards or 8 regular lanyards*	*160*
4	*auto locking carabiners*	*64*
Sub-total		*656*
TOTAL COST OF WALKWAY SYSTEM AND ACCESSORY EQUIPMENT		20,100

KEY: *DFG=drop forged galvanized   GV=galvanized*

Additional traverses to adjacent trees, consisting of two cables and a rope connecting two trees using a Bosun's chair and a trolley pulley with a lanyard, include $370 for the basic set-up equipment, plus chair ($150), extra ascender ($45), trolley pulley ($85), special lanyard ($50), extra carabiner ($30) and 7/16" galvanized screw shackle ($10). Cables and attachment hardware cost approximately $180.00 plus $2.36 per foot of cables and rope plus labor of approximately $300 for an average span. Such traverses could also be set up on a temporary basis using industrial slings made of nylon or polyester webbing, which would cost approximately $130 for the slings and hardware, but only $200 for labor.

This sort of arrangement allows the researcher to sit in the Bosun's chair, tethered to both the upper and lower cables as well as the center rope, while walking to any place on the traverse. Such traverse systems can be conveniently set up in spans up to 30 m in length.

2

Temperate Canopy Cranes

Basel Canopy Crane, Switzerland

By Christian
Körner

Location:
Hofstetten, NW
Switzerland.
Length of
operation: 3 years.
Grid reference:
47°28'N, 7°30'E.

THE Swiss Canopy Crane was built by helicopter in March 1999. It is situated in a fenced research site (forest plot size ca. 1 ha), which is equipped with field laboratory, power and telephone lines (see Figure 1). The site is easily accessible from the city of Basel (ca. 20 min drive) and a forest road ends 80 m from the crane site.

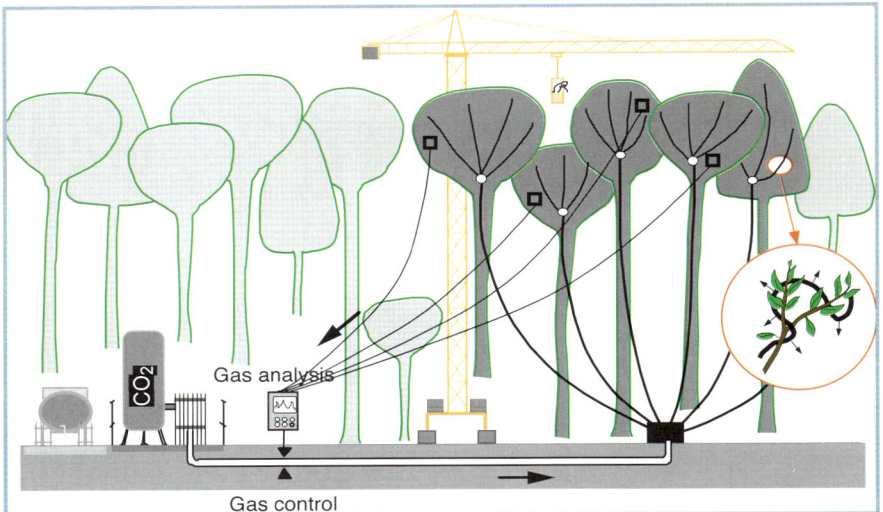

Gas analysis

Gas control

Figure 1 Swiss Canopy Crane Project.

Choosing a location

The criteria which dictated the choice of location were:
- a suitable site for the comparative study of functional ecology in mature European forest tree species at low altitudes
- as high a number of tree species present as possible
- representation of the major European forest tree species by as many individuals as possible
- as few logistical constraints to access and work as possible

The site is a typical central European low-altitude mixed forest at an altitude of 550 m.

Forest description

Climatically the site is in the humid temperate zone of western Europe, characterized by mild winters, moderately warm summers, a total annual precipitation of 800–1000 mm and annual evapotranspiration averaging around 550 mm. The soil is a silty loam with a depth of ca. 30 cm underlain by calcareous debris and rock with an average pH in the top 10 cm of the profile of 5.8.

The forest is about 120 years old (a secondary forest established on formerly marginal crop land), with tree heights between 32 and 38 m, and a total basal area of 46 m^2 ha-1 (Table 1). The stand has a leaf area index (LAI) of ca. 5 and is characterized by a dominance of European beech (*Fagus sylvatica*) and oak (both *Quercus petraea* and *Quercus robur* and their hybrids; the species are very similar, though the interspecific phenotypic variation exceeds that between the two species, hence trees are treated as *Quercus ssp.*) with subdominant representatives of lime (*Tilia platyphyllos*), hornbeam (*Carpinus betulus*), maple (*Acer campestre*), wild cherry (*Prunus avium*), as well as the four coniferous species European larch (*Larix decidua*), Norway spruce (*Picea abies*), silver fir (*Abies alba*), and Scots pine (*Pinus sylvestris*) (see Figure 2). Of these conifers, only *Abies alba* is regenerating under current conditions, while the other three species became established in the late 19th century (they are more light-demanding species). The spatial distribution of the ca. 70 trees (BHD >10 cm) within the crane's perimeter is shown in Figure 2.

Table 1 Stand characteristics by species in 1999

TREE SPECIES	STEMS PER HA	BASAL AREA (M2 HA−1)
Abies alba	57	1.26
Acer campestre	18	1.22
Acer pseudoplatanus	35	0.11
Carpinus betulus	64	2.94
Corylus avellana	11	0.02
Fagus sylvatica	106	11.02
Larix decidua	42	8.71
Picea abies	67	5.82
Pinus silvestris	21	3.26
Prunus avium	7	1.13
Quercus robur	67	8.46
Tilia sp.	117	2.37
TOTAL	615	46.31

In addition, the site has a strong presence of evergreen, climbing ivy (*Hedera helix*) reaching the canopy with arm-thick stems. There is a rich understory shrub flora with species like hazel (*Corylus avellana*), honeysuckle (*Lonicera xylosteum*), spurge laurel (*Daphne laureola*), and the evergreen holly (*Ilex aquifolium*). The herbaceous layer is dominated by Mercurialis perennis; other important species are *Paris*

quadrifolia, *Anemone nemorosa*, and *Galium odoratum*. On a hectare basis, the number of tree species is higher, since in the immediate vicinity of the crane circle, there is also ash (*Fraxinus excelsior*), whitebeam (*Sorbus aria*), and a second species of maple (*Acer pseudoplatanus*). Hence, including *Hedera*, 16 woody species reach the >30 m canopy height.

The crane was set up by a special two-rotor helicopter operating on a 100 m rope in March, when the deciduous trees were leafless. The whole procedure was completed within five hours. No building machinery/vehicles were allowed to enter the forest. The few concrete sockets for the crane were made by hand. More than 90% of the time we use the small round gondola, because it creates much less damage to the trees and is far easier to manoeuvre (Figures 3 and 4). The crane was sponsored by the Swiss Federal Office of the Environment (BUWAL) and the University of Basel.

Figure 2
Swiss Canopy
Crane: tree
positions.

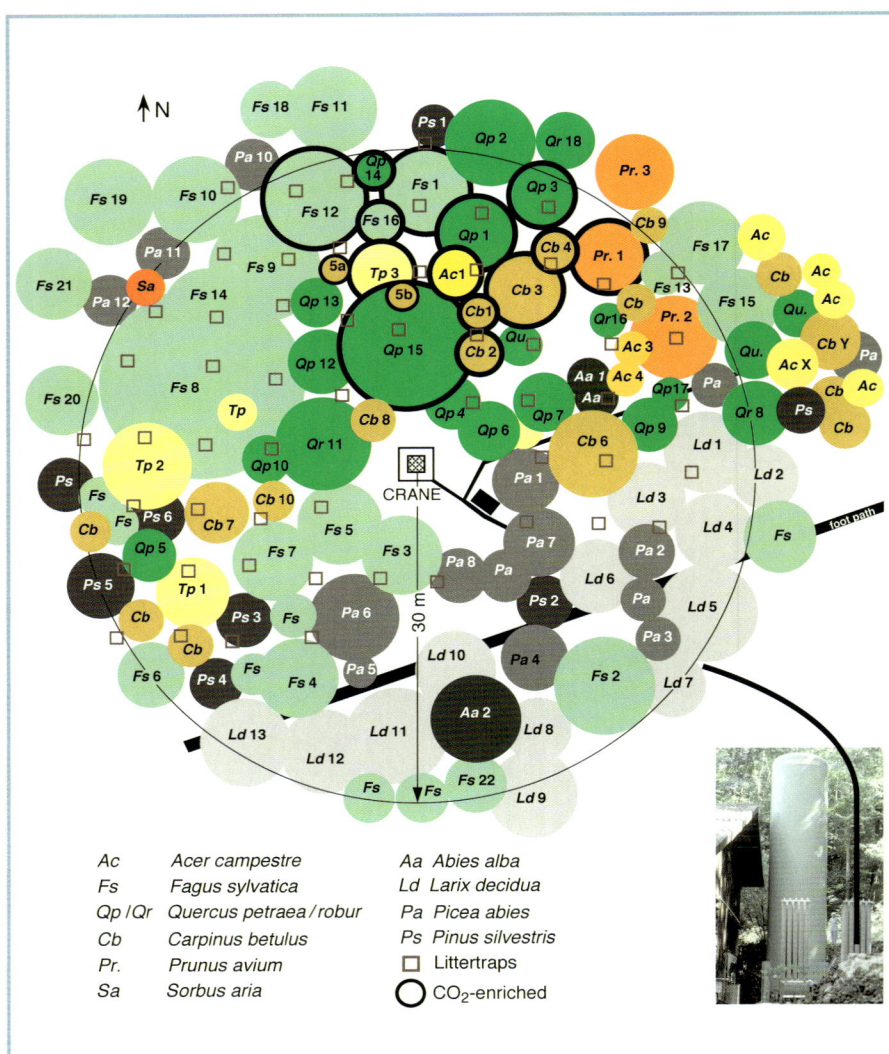

Ac	Acer campestre	Aa	Abies alba
Fs	Fagus sylvatica	Ld	Larix decidua
Qp/Qr	Quercus petraea/robur	Pa	Picea abies
Cb	Carpinus betulus	Ps	Pinus silvestris
Pr.	Prunus avium	☐	Littertraps
Sa	Sorbus aria	◯	CO_2-enriched

Table 2 Crane vital statistics

Make	Liebherr 30LC with 8 m customised basal tower extension
Fixed/mobile	Fixed on 4 separate concrete sockets
Tower height/hook height	45 m/39.5 m
Jib length	30 m
Max. height reached with gondola	Circa 37 m
Gondola type	Both designed by Lifting Technologies Inc., Montana, USA
	A. round design Model RM1-300A/32; 65 cm diameter
	B. square design for bulky instrumentation Model SEC 02/600, 1.2x1.2 m
Number of people recommended	A. 1 person; B. 4 persons
Area forest accessed	0.28 ha*

*Note, this area is fully forested, the crane itself is situated in a small natural gap of only a few m²

Research projects

In these days of intensive forest management of just a few commercially valuable species, the site's species diversity represents a rather unique situation. The major European forest tree species are represented in substantial replication within a small area. This offers the possibility of comparative studies in plant sciences, entomology, forest pathology and others, and we invite all partners to capitalize on this opportunity. In other words, we do not encourage single tree species studies at this site. We believe that the site should assist in arriving at a better functional understanding of forest biodiversity in central Europe.

One of the major research activities at this site is canopy CO_2-enrichment using a new technique called web-FACE (Pepin & Körner, 2002). We release pure CO_2 in thin tubes woven into the canopy (2 tonnes per day for 14 trees). We have completed a first full season of tall forest tree exposure to 510 ppm of CO_2 in 2001. This project aims at finding tree species-specific responses to CO_2-enrichment. We see reductions in stomatal conductance of around 15% in deciduous species (no effect was seen in conifers using the branch bag technique in 1999). Sap flux data from 2001 suggest a reduction of transpiration of ca. 10% due to elevated CO_2. Other research teams have examined isopren emission, insect abundance and leaf chemical composition. A major effort is stable carbon isotope work. 13C signals imposed by CO_2 enrichment are followed from tree canopy down to mycorrhizal fungi.

This is the first ever attempt at simulating a

Figure 3 Gondola being lowered from the crane jib. Photo: Christian Körner

future CO_2-rich atmosphere in the canopy of a mature natural forest, one of the urgent needs, to understand forest responses to global change (Körner 2000). The Swiss Canopy Crane is managed by the Institute of Botany, University of Basel (website: www.unibas.ch/ botschoen or search the internet for "Swiss canopy crane").

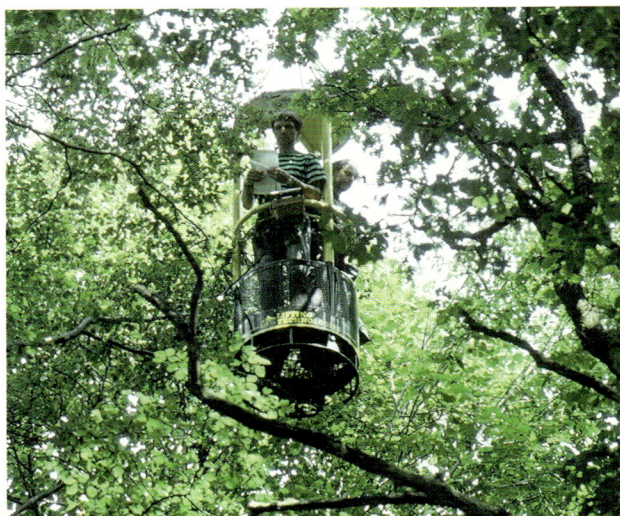

Figure 4 Researchers in the gondola in the canopy. Photo: Christian Körner

Table 3 Costs

EXPENSE – ONE OFF	ESTIMATED COSTS IN US$
Second hand crane	148,000
Crane construction with helicopter	12,000
Power supply (70 Amps, in our case a 1 km new power line)	24,700
Concrete foundation and trails	12,300
Workshed and miscellaneous (2 x 1.2 m)	6,200
Fence around the crane area	3,000
Crane driver license (courses)	3,000
TOTAL	209,200
ANNUAL COSTS	
Annual service and power consumption	2,000
Site rental	1,200
Crane operators and mechanics (part time)	21,600
TOTAL	24,800

Contact details

Director: Prof. Christian Körner (email: Ch.Koerner@unibas.ch)
Research associate: Dr. Gerhard Zotz (email: gerhard.zotz@unibas.ch)
Technical experts: Alain Studer and Luca Egli (both MSc in physics)
Chemical laboratory and sampling: Olivier Bignucolo

Literature sited

Körner Ch (2000). Biosphere responses to CO_2-enrichment. Ecol Appl 10(6):1590–1619
Körner Ch (2000). Baumkronenforschung: abgehoben oder von hoher Warte? Gaia 9:165–169
Pepin S & Körner Ch (2002). web-FACE: a new canopy free-air CO_2 enrichment system for tall trees in mature forests. Submitted to Oecologia

Freising Canopy Crane, Germany

By Rainer
Matyssek and
arl-Heinz Häberle

Location:
"Kranzberg
Forest", Freising,
Bavaria, Germany.
Length of
operation: 1 year.
Grid reference:
11°39'41"E,
48°25'08"N.

THE fenced research plot is 0.5 hectare in size at an elevation of 485m and comprises about 500 spruce (*Picea abies*) and beech (*Fagus sylvatica*) trees in a managed forest that forms a closed canopy (Pretzsch *et al* 1998). The trees are between 50 and 60 years old, and the foliage extends from 17 m above ground to the upper canopy edge at 28 m.

Crane vital statistics

At this site we use two methods to access the canopy, the first of these is a tower crane (KROCO) details of which are given below:

Table 1

Make	*Potain (France)*
Fixed/mobile	*Fixed*
Tower height	*42 m*
Jib length	*50 m*
Max. height reached with gondola	*35 m*
Gondola type	*cylindrical 0.7 m diameter*
Number of people carried	*Max. 2*
Area forest accessed	*0.8 ha*

The second of these is four scaffolding towers, ranging between 27 and 35 m in height. Three towers are connected by platforms (12 m in length) that allow access at four heights between 17 and 25 m to the sun and shade crowns of about 30 trees (installed in December 1996).

Research projects

The main research aims pursued are part of the interdisciplinary research program "Sonderforschungsbereich (SFB) 607"

1. Quantification of competitive interactions between adult beech and spruce trees based on the resource fluxes involved;
2. Clarification of the associated resource allocation, employing experimental 'free-air' ozone (O_3) exposure within the canopy to facilitate the assessment of regulatory mechanisms and their responsiveness (see below);
3. Determination of eco-physiologically meaningful threshold levels in the O_3 sensitivity of adult forest trees and mitigation of their carbon sink strength under enhanced, chronic O_3 exposure (see below).

Experimental free-air canopy O_3 exposure

In canopy research at Kranzberg, ozone plays a dominant role in two ways: first as a tool to study the regulation of carbon allocation and its importance for the competitive mechanisms existing between beech and spruce, and second as an

air pollutant with the highest toxic potential in Europe which appears in the course of 'summer smogs'.

For both these aspects information from adult forest trees is rather scarce and cannot be gained without access to the tree crowns. Consequentially a 'free air' ozone exposure system has been installed (see Figure 1)(Werner & Fabian 2002). Within a volume of 2000 m^3 which comprises the crowns of 10 neighboring trees, the O_3 levels prevailing at the forest site are increased to "2 x ambient level" (up to a maximum of 150 ppb O_3). The enhanced O_3 levels are released through a system of 130 teflon tubes (outlet capillaries at 3-dimensional distances of about 0.4 m, compensation for internal pressure gradients) which vertically extend across the entire foliated canopy. Five spruce and five beech trees are exposed to this regime, while another group of five trees each growing under the ambient O_3 regime (i.e. unchanged O_3 levels) serves as a 'control' (which is usually unavailable in outdoor O_3 studies). This methodology ensures an experimentally enhanced and chronic whole-tree exposure to ozone while avoiding – in the absence of plant enclosure in chambers/cuvettes – physiological bias through micro-climatic artifacts (as prevailing in conventional fumigation studies). The striking advantage of this system is its applicability to adult trees growing in natural forest stands (Karnosky *et al* 2001). The free-air O_3 fumigation has been in operation since May 2000.

The equipment required for the generation and control of ozone and for the assessment of environmental factors is an ecophysiological field laboratory (for analysis of leaf gas exchange, xylem sap flow, stem, branch, root and soil respiration, growth rates and biomass production), and a central laboratory hut as part of the infrastructure of the experimental site.

Experimental concept

The concept focuses on the functional performance of individual trees growing in stands with multi-factorial biotic and abiotic interactions. With respect to research aim (3) above, the comparison between the two O_3 regimes enables quantitative risk assessment of broad spectrum molecular, biochemical and ecophysiological tree responses under prevalent site conditions. This allows the examination of tree processes which are at risk or already reflect incipient injury under the atmospheric O_3 regime. Each of the study trees is statistically viewed as an individual case study, which aims to derive consistent patterns from synchronously occurring O_3 responses at the cell, organ and whole-tree level. Mechanistic modelling is used to scale-up findings to the stand level. The response patterns form the basis for comparisons between the two O_3 regimes. The experimental design represents an appropriate statistical approach, considering the obvious logistical restrictions of extending the fumigation methodology to several experimental plots. The findings generated will help define field-relevant measures of O_3 sensitivity in adult forest trees which do not exist at the moment. They will also make thresholds based on O_3 exposure obsolete while promoting concepts of actual O_3 uptake, i.e. the O_3 flux into leaves through stomata. Through this method physiologically and ecologically meaningful O_3 doses are provided with respect to tree acclimation to ozone, changes in the carbon sink strength of forests, and initiation of O_3 injury (Häberle *et al* 1999).

Measuring actual O_3 uptake is fundamental to pursue research aims (1) and (2) of SFB 607 (above), i.e. the quantification of competitiveness based on resource flux. Competitiveness is assessed as a hierarchy of efficiency ratios in space sequestration (resource investment per unit space occupied above and below ground), resource acquisition (resource gain per investment and occupied space), and 'running costs' (transpiration, respiration per resource gain and occupied space). In this way, 'cost/benefit' relationships are established in the control of resource allocation within and amongst the trees being exposed to biotic and abiotic impacts (Grams *et al* 2002). Here, ozone is used as a disturbant rather than pollutant to unravel the responsiveness and differentiation of such relationships as well as their underlying mechanisms.

Another fundamental factor being studied is the light gradients which occur naturally across the canopy and their influence on such 'cost/benefit' relationships. In addition, the impact of pathogens, mycorrhizospheric organisms and phytophagous insects is of paramount importance with respect to the rationale of SFB 607 (Matyssek *et al* 2002). Thus impact levels of parasites and their effects on trees' primary and secondary metabolism are being assessed, and that sampling is performed on the abundance of insects in the canopy of the stand.

The investigations outlined above take advantage of the broad scope of expertise that is brought in by the numerous working groups integrated into SFB 607. 20 projects are cross-linked with each other, covering the fields of molecular biology, biochemistry, ecophysiology, plant nutrition, mycorrhizae, phytopathology, zoology, soil science, forestry, air chemistry and modelling. About 10 of these projects are integrated into the scientific activities at Kranzberg Forest.

Added to the partners in SFB 607, collaboration exists with about 10 working groups, both at the national and international level. There is interest in extending cooperation with external groups that may contribute in a complementary way to the research conducted at Kranzberg Forest. For further information see www.forst.tu-muenchen.de/EXT/SFB607/sfb_607.htm

Costs

There was a total budget of 175 000 Euros for the purchase and construction of the crane. Half was used to purchase the crane and the other half covered construction and power supply.

Funding

This is being provided for research projects (1) and (2) by "*Deutsche Forschungsgemeinschaft*" (DFG) through the interdisciplinary research program "Sonderforschungsbereich (SFB) 607: *Growth and Parasite Defense – Competition for Resources in Economic Plants from Agronomy and Forestry*" (since

Figure 1 View of the scaffolding with the 'free air' ozone fumigation system; from the canopy crane in April 2001 before bud break. Photo: Häberle.

1998), and for (3) by "*Bayerisches Staatsministerium für Landesentwicklung und Umweltfragen*" (Bavarian Ministry of Environment) through the project "*Risk Assessment of the Enhanced Chronic O_3 Exposure by Means of "Free-Air" Canopy Fumigation in a Mixed Beech/Spruce Forest*" (since 2000; coordinator of both projects R. Matyssek). At present, funding is ensured through 2004, and it is intended to extend the research program by at least a further 3-year period. Routine maintenance of the crane is funded by the University of Munich and the land it stands on is provided 'free of charge' by the Bavarian Ministry of Agriculture and Forestry.

Contact details

Dr. Rainer Matyssek & Dr. Karl-Heinz Häberle, Dept. of Ecology / Ecophysiology of Plants,
Technische Universität München, D – 85354 Freising-Weihenstephan, Germany
email matyssek@bot.forst.tu-muenchen.de.

Literature cited

Grams TEE, Kozovits AR, Reiter IM, Winkler JB, Sommerkorn M, Blaschke H, Häberle K-H & Matyssek R (2002). Quantifying Competitiveness In Woody Plants. *Plant. Biol.* (In Press)

Häberle K-H, Werner H, Fabian P, Pretzsch H, Reiter I & Matyssek R (1999). 'Free-Air' Ozone Fumigation Of Mature Forest Trees: A Concept For Validating AOT40 Under Stand Conditions. In: Fuhrer J & Achermann B (Eds.) *Critical Level For Ozone – Level II.* Swiss Agency For The Environment, Forests And Landscape (SAEFL), Berne, Pp. 133–137

Karnosky DF, Gielen B, Ceulemans R, Schlesinger WH, Norby RJ, Oksanen E, Matyssek R & Hendrey GR (2001). FACE Systems For Studying The Impacts Of Greenhouse Gases On Forest Ecosystems. In: Karnosky DF, Scarascia-Mugnozza G, Ceulemans R & Innes JL (Eds.) *The Impacts Of Carbon Dioxide And Other Greenhouse Gases On Forest Ecosystems.* CABI Press, Pp. 297–324

Matyssek R, Schnyder H, Elstner E-F, Munch J-C, Pretzsch H, Sandermann H (2002). Growth And Parasite Defense In Plants; The Balance Between Resource Sequestration And Retention. *Plant. Biol.* (In Press)

Pretzsch *et al* (1998). Die Fichten-Buchen-Mischbestände Des Sonderforschungsbereiches "Wachstum Oder Parasitenabwehr?" Im Kranzberger *Forst. Forstw. Cbl.* 117: 241–257.

Werner H, Fabian P (2002). Free-Air Fumigation Of Mature Trees. *Environ Sci & Pollut Res* (In Press)

Hokkaido Canopy Crane, Japan

I N 1997, a canopy crane was installed in a deciduous broad-leaved forest close to the Pacific coast, in Hokkaido island, northern Japan.

By Masashi
Murakami and
Tsutom Hiura

Location:
Tomakomai
Experimental
Forest (TOEF),
Hokkaido island,
Japan.
Length of
operation: 5 years.

Choosing a location

The criteria which dictated the choice of location were:
• need for well prepared background data
• on site facilities for canopy researches
• a primary forest site
• electricity supply
• an area flat enough for construction

The crane is located in an area of natural forest near the town of Tomakomai. This 330 year old deciduous broad-leaved forest receives on average 1200 mm of water-fall per year and has a four-month snow season from December through March. TOEF (2715 ha) is composed of primary, secondary and plantation forests. One third of the area is secondary forest which has developed after a typhoon hit the area in 1954, one third is mature forest, and the final third is plantations. The site is about 10 minutes' drive from the headquarters of the experimental forest.

The crane is able to reach 21 species of trees and 6 species of lianas within the forest. It stands in an experimental plot of 9 ha in which all individuals of 10 cm diameter at breast height (dbh) or greater have been measured, mapped and iden-tified. A large scaffolding unit (16 x 16 x 22m) was also established in the 9 ha plot to investigate three-dimensional structure and function of the forest at a patch scale.

The crane was constructed by the Konoike Construction Company (Japan) and took about two months in total. The main parts were installed by a mobile crane after pre-assembly outside the forest. Fortunately, the construction process did not damage very much of the forest around the crane (only two trees > 10 cm dbh were cut), this was due to T. Hiura explaining over and over again about the importance of the forest to local contractors and forest labourers. Maintenance of

Table 1 Crane Vital Statistics

Make	Ogawa Industries, Model OTH-80N
Fixed/mobile	Fixed on a concrete anchor and base
Tower height	25 m
Jib length	41 m
Max. height reached with gondola	Circa 20 m
Gondola type	A: cylindrical 2.1 x 0.7 m; B: cylindrical 2.4 m x 1.2 m
Number of people carried	A: max: 1 person; B. max: 5 people
Area forest accessed	0.5 ha

the crane requires 15 working days of two experts per year, which costs approximately $10,000. Bimonthly inspection is conducted by a local company, and experts from the supplier (Sapporo) check the crane once a year. Besides this, safety tools are checked daily by crane users and operators.

Research projects

TOEF is dedicated to research into terrestrial environments in the temperate regions of the world. TOEF hosts 150 visiting scientists each year and provides facilities for their work, in the form of laboratories, housing, libraries, and research permits.

To date the projects listed below have been carried out using the canopy crane. Biodiversity:
- Mechanism of masting related to storage reserves in different tree species. We propose to measure stored reserves in trunks (0.3 m, 1.3 m, 4 m diameter) and branches of trees, mostly by microscope observation. Sampling is also scheduled to include various phenological stages spread over several years.
- Relationships between the three-dimensional structure of the forest and insect diversity. Microdistribution of insects, mainly drosophilids, flying in the above-ground space of the forest was surveyed on a small scale, in relation to three-dimensional foliage distribution.
- Spatio-temporal variation in moth larval communities dependent on oak. Moth larvae were collected from sunlit and shaded leaves of canopy oak trees. Leaves were sampled by hand in the spring and summer, and measurements of several physical properties of the habitat and physio-chemical properties of the leaves were made.

Climate change
- Daily, seasonal and annual change in leaf-level photosynthetic characteristics as related to successional traits. To understand the factors affecting daily and seasonal changes in photosynthesis, we simultaneously measured a suite of environmental conditions surrounding the leaf to quantify species-specific sensitivity to environmental conditions as they change throughout the growing season.
- Carbon and nitrogen dynamics of a deciduous broad-leaved forest. Seasonal changes in carbon and nitrogen dynamics within the forest were estimated as a function of nitrogen and carbon flow through plants.

Costs

The costs of construction and purchase of the crane were about $670,000 in total, including $17,000 for construction (mainly electronic devices). The cost of maintenance changes in accordance with the number of repairs which ranges between $10,000 to $17,000 per year. Another $2,5000/year is needed for electric charges.

Funding

This canopy crane was funded by the Ministry of Education, Science, Sports and Culture of Japan through a program called "An integrated study into biodiversity conservation under global climate change and a bio-inventory management system". Tomakomai Research Station manages the crane and decides on the projects that are to be allowed at the site.

Contact details

Dr. Masashi Murakami, Tomakomai Research Station, Field Science Center for Northern Biosphere,
Hokkaido University, 053-0035, Japan.
Tel: +81-144-33-2171, Fax: +81-144-33-2173
Email: masa@exfor.agr.hokudai.ac.jp

Leipzig Canopy Crane, Germany

By Wilfried
Morawetz and
Peter Horchler

Location: Leipzig,
Germany.
Length of
operation: 1 year.
Grid reference:
51°20'16"N,
12° 22'26"E.

PROF Morawetz decided in March 2001 to set up a crane system in the north-western part of the extensive floodplains around the city of Leipzig (Germany). He had previously built a tower crane running on a rail track in the Suromoni forest of Venezuela and had learnt a lot from this project (see Venezuela crane later in this Handbook).

Choosing a location

The criteria which dictated the choice of location were:
- a Central European deciduous forest in which detailed investigations could be carried out
- existence of good baseline information on biotic and environmental data
- highly diverse and nearly natural species composition and stand structure
- minimal logistic problems reaching the location even by public transport

Three alternative locations within the Leipzig floodplain forest area were taken into consideration. The decision for the actual location, taken in accordance with the city's mayor and forestry administration, was based on (1) the presence of a dyke on which the railtrack could be built, thus causing only minor disturbance of the forest and (2) the presence of a building which could be used as field camp. Although the area where the crane is set up is a nature sanctuary, there were no problems with the environmental administration or with NGOs. Nevertheless it took about one year to get through all the administrative procedures to get approval for installation of the crane. The steps were:
- finding the location in co-operation with the forestry administration
- getting expert opinion on ground stability
- receiving approval from the city's state environmental administration
- getting agreement from NGOs
- drawing up a land rental contract between the university and the city

The choice of the Liebherr crane was based on previous positive experience with

Table 1

Make	Liebherr 71EC
Fixed/mobile	Mobile on a 120 m rail track
Tower height	40 m
Jib length	45 m
Max. height reached with gondola	Circa 33 m
Gondola type	Used for window cleaning on skyscrapers, rectangular, wt. 250 kg
Number of people carried	Max. 3
Area forest accessed	1.6 ha

Crane vital statistics

the same equipment in the Surumoni project (Venezuela). Added to this the purchase of the used crane and all restoration work needed was sponsored by the Liebherr company to whom we are very grateful.

This crane is unusual because it was built on a rail track which facilitated access to a much larger area of forest (see Figure 1). Figures 2 and 3 illustrate the tracks and engine used to move the crane along the transect. The first step was stabilisation of the existing earth dyke in order to carry the cranes' weight. After building a short stretch of the rail track, the cranes' tower was erected by means of a mobile crane truck. The remaining rail track was then built using the tower crane moving along the tracks that had just been laid down. Preparation for this construction work involved felling 16 trees and removing some branches from another 15 in order to facilitate movement of the crane tower.

Figure 1 Crane on rail tracks. Photo: Wilfried Morawetz.

Research projects

The crane has only recently been established and thus there has only been time for some short-term studies looking at canopy arthropods, bats, macro fungi, bryophytes, lichens and some aerial algae.

Some major research projects will start in 2002. These will comprise detailed studies of the groups already mentioned, focussing especially on canopy arthro-

Figure 2 Close up of the tracks (left). Photo: Andrew Mitchell.
Figure 3 Close up of power cable drum and small electric motor used to move the crane (above). Photo: Andrew Mitchell.

pods, as well as studies of the phenology (see Figure 2), reproductive ecology and population genetics of the dominant tree species (*Fraxinus excelsior* and *Quercus robur*). Data loggers are also to be installed above and within the canopy in order to gather basic microclimatological data. This multi-disciplinary research can be subdivided into four main areas, biodiversity, biological processes, climate and soil.

The key objectives (especially for biodiversity) are to find answers to the following basic questions:
- What's up there? (Inventory of canopy organisms, canopy processes)
- Is there a spatial structure i.e. vertical and horizontal differentiation of canopy organisms, microclimate and biological processes?
- Is there temporal structure of canopy organisms, climate and biological processes?
- What's the function of these structures within the ecosystem?
- What is the applied environmental outcome of our findings?

Further details concerning the proposed research project will soon be provided on the projects' web page (www.uni-leipzig.de/~instbota/LAK.htm).

All data gathered in the project will be entered into a data base and linked to a Geographical information centre (GIS) (ARC-Info) in order to provide reference data for all project participants. Our goal is to have frequent meetings, with a number of the project participants, in order to provide a continuous and complication-free flow of information.

So far, co-operation has been planned for 2002 with some members of the Kranzberg Ozone Canopy Crane Project (KROCO) recently installed near Freysing (Southern Germany) by the Technical University Munich, and the canopy scientists from the University of Würzburg. Further co-operation is planned, especially with the other canopy crane projects in temperate climates. These are the Swiss Canopy Crane Project, Tomakomai Experimental Forest (Japan), and the Wind River Canopy Crane Facility.

Costs

So far we do not have enough hard facts concerning all the annual project running costs. The following table gives rough estimates:

Figure 4 Sampling a cherry tree from the gondola. Photo: Punctum

Table 2

EXPENSE – ONE OFF	ESTIMATED COSTS IN US$	TIME TAKEN TO BUILD
Rail track	30,000	*1 month*
*Crane purchase**	28,000	*1 week*
*Dismantling of the crane and rail track***	30,000	
TOTAL	88,000	
ANNUAL COSTS		
1 Co-ordinator	47,000	
Energy	1,000	
Ongoing smaller maintenance work	200	*7 days per year*
Big maintenance (annual)	600	*1–2 days per year*
Minor repair work	200	
Transportation to the facility	400	
TOTAL	49,400	

Costs for a used and restored crane, a new crane costs about US$ 200,000

**According to the contract with the city Leipzig the area where the crane was set up is rented for ten years. After this period the crane and the rail track has to be dismantled unless the project continues.*

Funding

Initial three-year funding (ca. US$ 300,000) including the set up and maintenance of the crane, as well as the personnel costs for the project co-ordinator has been provided by the Centre for Environmental Research Leipzig-Hallé (UFZ). Proposals for further funding are underway. The crane administration itself does not fund any projects.

In order to buffer some of the running costs, a fee of Euro 25 (ca US$ 22) per hour is charged to all scientists using the crane.

Pros and cons of the canopy crane technique

A crane provides easy-to-use and almost stress free access to most parts of the canopy in a fairly large area (1.6 ha). Especially important is the access offered to outer parts (twigs) of tree crowns. However, in certain circumstances if the scientist needs to reach branches "deep" in the crown the gondola can not be manoeuvred and it is necessary to use ropes to climb down from it.

Using a crane is especially suitable for long term studies in a restricted area which need easy and frequent access. The major constraint is the limited area of the canopy which can be accessed. Hence, selection of the site has to be made very carefully. Initial costs are also relatively high but running and maintenance costs are low. Use of a crane is not suitable for studies which are large scale or are temporarily restricted.

Contact details

Prof. Dr. Wilfried Morawetz/Peter Horchler, Universität Leipzig, Institut für Botanik, Spezielle Botanik & Botanischer Garten, Johannisallee 21-23, 04103 Leipzig, Germany Email: horchler@uni-leipzig.de, Tel: ++49-341-9738590, fax: ++49-341-9738549

Solling Canopy Crane, Germany

By Michael
Bredemeier,
Norbert
Lamersdorf,
and Gustav
Wiedey

Location: Solling
mountains, central
Germany.
Length of
operation: Ten
years.
Grid reference:
51°31'N 9°34'E.

Choosing a location

The crane research facility was sited at Solling because the site has been a focus of intensive interdisciplinary forest ecosystems research since the 1960s. The site was part of the Man and the Biosphere Programme (MAB) and the International Biosphere Programme (IBP). This meant that data from former monitoring and research projects was available to be used as baseline information.

Crane vital statistics

One of the main factors which dictated the use of a tower crane was the need to access the crowns of all spruce trees in the area of the roof manipulation facility, plus the system had to be relatively inexpensive and low-maintenance.

Permission to develop the research facility had to be obtained, but this proved to be relatively straight forward. It was not necessary to get clearance from the government, but only from the local council and the university itself.

The crane was erected in the central area between the three roofs (see Figure 1) by employing an 80T mobile auto-crane, which could lift each of the segments into place. The whole process took one week, of which two days were needed for construction of the crane itself.

The gondola is operated either from within or by remote control from the ground. It was supplied by the Greifzug company. Overall technical planning was delivered by NTT Neuhaus Transtech GmbH.

Table 1

Make	Liebherr modified
Fixed/mobile	Fixed on a concrete foundation
Tower height	33 m
Jib length	25 m
Max. height reached with gondola	Circa 28 m
Gondola type	1.5 m x 0.7m x 1.2 m high, purpose built, carrying capacity 500 kg
Number of people carried	Max. 2
Area forest accessed	0.2 ha

Liebherr-Werk Biberach GMBH, P.O. Box 1663, D-7950 Biberach an der Riess 1, Phone (07351) 41-0, Fax.: (07351) 41225

Greifzug Hebezeugbau GmbH, P.O. Box 200440, D-Bergisch Gladbach 2, Phone (02202) 1004-0, Fax.: (02202) 1004-50

NTT Neuhaus Transtech GmbH, Dorfstr. 8, D-8155 Valley-Oberlaindern, Phone (08024) 49900, Fax.: (08024) 49444

Research projects

Before construction of the crane some data was gathered from the site, including information on soil mechanical properties and local forest structure.

Research undertaken at this site mainly concentrates on tree-physiological measurements (short-term), tree growth measurements and sampling of tree compartments in the canopy. Some baseline information has been collected which includes rates of forest growth and seasonal transpiration rates.

Work has been carried out in cooperation within the European Union research networks NITREX and EXMAN. As yet there have been no strictly standardized protocols developed for research. New projects have to fit in with the overall interdisciplinary integrated research within the Solling roof project; they must also be approved by a coordination committee.

Figure 1 Schematic representation showing the crane above the roofed manipulation facility.

Costs

Experience has shown that the system could have been built even more cost-effectively and would have been more flexible if a passive gondola without a lift of its own had been used and operated together with the other crane functions completely by remote control.

Table 2

EXPENSE	ESTIMATED COST US$
Crane	13600
Crane modification	13600
Gondola	9100
Remote control	4500
Transport and construction on site	22700
Maintenance	2700
TOTAL	66000

Funding

The crane administration itself has no resources for funding available and merely acts as a facility operation and steering committee. The funding came from the German Federal Ministry of Research and Technology (BMFT).

Contact details

Dr. Michael Bredemeier, Dr. Norbert P. Lamersdorf, Dr. Gustav A. Wiedey; Forest Ecosystems Res. Ctr., Univ. of Goettingen; Buesgenweg 1, D-37077 Goettingen, FRG; Germany
email: mbredem@gwdg.de, nlamers@gwdg.de, gwiedey@gwdg.de

Wind River Canopy Crane, USA

By David Shaw

Location: Gifford
Pinchot National
Forest,
Washington State,
USA.
Length of
operation: 7 years.
Grid reference
Latitude: N 45°
49''13.76"
Longitude: W
121°57'06.88".

Choosing a location

The criteria which dictated the choice of location were.
• old-growth (300 yr +), tall stature forest
• flat topography, suitable for tower crane 360° swing
• multi-species canopy/forest with Douglas-fir present
• no un-natural edge, i.e. only interior forest
• comparable to other forests in the region

Many sites were looked at for construction of this crane. It was originally proposed for the western Olympic Peninsula, Washington State, from Forks to Quinault. The first choice was in the Olympic National Park, but this was unacceptable to the park authorities. The entire west side of the peninsula was then scouted using airplane and ground assessment. A site near Lake Quinault in the Quinault Rain Forests was finally chosen, but the permit application was turned down. The search was then re-located to the Cascade Mountains where 3 locations were considered in the Wind River Experimental Forest and 6 locations in the H.J. Andrews Experimental Forest (near Eugene, Oregon).

The procedures and legislation that had to be followed before installation of the crane were varied and complex involving the inclusion of many stakeholders (as detailed below).

Governmental. US Forest Service procedures were followed, requiring an Environmental Assessment (similar to an Environmental Impact Statement, but not as detailed). Approval was also required from engineers, wildlife biologists, hydrologists, fisheries biologists, rare plant specialists, and the Experimental Forest Managers.

Local Councils. Local government and public approval was required by the US Forest Service. We met with Chamber of Commerce, County Commissioners, County Sheriff, Economic Development Administration, and had several public meetings. US Forest Service facilitated the public approval process.

University. University appointed a Project Administrator from their construction shop whom I worked with to secure University oversight. He managed the purchasing, crane inspections, and construction process in collaboration with us. I was present on site for the entire process, whereas the Project Administrator was not.

Other. The crane is in a Research Natural Area (RNA). RNAs are set aside for the study of ecosystems in their natural state. Destructive sampling is discouraged. The crane projected required approval of the regional RNA oversight committee. The final choice of location was heavily affected by political considerations. The original proposal to site the facility on the western Olympic Peninsula was opposed by an organised group (the Washington Commercial Forest Action Committee) affiliated to timber cutting interests. Our permit application took 16 months to process and was finally turned down in spring of 1994 due to threats of violence.

The decision was then made that the facility must go to an existing experimental forest managed by the US Forest Service. The Washington site was chosen based on forest characteristics and logistics of access (no new road was required). The crane is accessed by walking about 1 km from our office or 580 m from a gate down the access road. Alternatively one can take an electric golf cart to the site for those with lots of equipment.

The crane was supplied by the Morrow Crane Company, Inc. Salem, Oregon, USA. Minimum requirements for the bid followed university protocols. The original company that won the bid was Pecco USA, who proposed a used (10 yr old) Pecco 440 for the site. Pecco was purchased by Morrow soon after we accepted their bid and the engineers at Morrow/Liebherr Germany thought the Pecco would not be appropriate. They proposed we substitute a used (10 yr old) Liebherr 550 HC instead, which is a bigger and more robust crane, although the motor system is not as strong as most tower cranes. We were thrilled, as the Liebherr crane is much better and we feel we got a great deal.

The crane was most recently used to build the Public Library in downtown San Francisco (1993/94). So we are fond of saying, "This crane went from building a citadel of knowledge to becoming a fountain of knowledge!"

The construction sequence used for installation of the crane was as follows:
1. Hazard tree and snag survey
2. Felling hazard snags
3. Upgrading road

Figure 1 Looking down the tower from the gondola. Note the people on the ground, with hard hats and the small footprint from the construction activity and installation of the crane.
Photo: University of Washington

4. Foundation of crane built
5. Power line installed
6. Crane erected

Table 1 Crane vital statistics

Make	Liebherr 550 EC
Fixed/mobile	Fixed
Tower height	74.5 m
Jib length	85m
Max. height reached with gondola	Circa 67 m
Gondola type	A. Floor = 1.2 m x 1.2 m. Cage height: 2.2 m. Max load 454 Kg
	B. Floor = 2.7 m x 1.3m. Cage height = 2.2 m. Max load = 908 Kg
Number of people carried	A. max. 4 persons; B. max. 8 persons
Area forest accessed	2.3 ha

The crane was installed adjacent to an old, logging road that had been closed since 1984. The road was upgraded to the crane site (geofabric laid down and gravel put on top of that). Electrical power (440v 3-phase) was tied into existing power at the Wind River Nursery, and pulled for about 1 km to the site. The power cable was buried in the middle of the road. The crane foundation was placed adjacent to the road, and a mobile crane was used to erect the tower crane.

We feel quite lucky to have had great working relationships with the US Forest Service, University of Washington, Morrow Crane Company, and the contractor that developed the site and erected the crane. Most important was to have our

Figure 2 View from helicopter. Located in a 500 year old forest dominated by Douglas-fir (*Pseudotsuga menziesii*), western hemlock (*Tsuga heterophylla*), and western redcedar (*Thuja plicata*). Photo: Jerry Franklin

research manager on site during the site development and construction process, keeps everyone honest and they related well to our mission. I learned how to swear well, and drink with the best of 'em.

Research projects

Prior to the erection of the crane maximum tree heights were determined. A 4 ha stem mapped plot was also established in summer 1994, and the crane was then placed near center of this plot. Data measured included diameter, species and location of all trees >/= 5cm diameter at breast height. Basic data on plant species, lichens, moss, and liverworts were also collected (plot is now 12 ha).

Baseline information is being collected at the site. This comprises species lists of flora and fauna including lichens, mosses, fungi, birds and vascular plants. Mammals and reptiles/amphibians are listed from what is expected to use the area.

In the 12 ha plot surrounding the crane the stems of all trees >/= 5 cm dbh have been mapped and are measured every five years for growth and mortality. Litter fall is recorded via monthly collections from 20 litterfall collectors positioned in the forest. These can be used for nutrient cycling as well. The water table is also measured weekly with four Peizometers placed in wells across the site.

The facility has its own weather stations: 1 in an open field near the site, 6 vertical stations located on the tower crane and 1 near the base of the crane. Together these provide fairly comprehensive weather statistics.

Research has been conducted under four broad functional areas, these are:
1. Organism biology/ecology
2. Process ecology
3. Tree ecophysiology
4. Global climate change, and atmospheric science

A canopy crane facility is good for studying the following things:
- tree physiology studies, because one can get replication in the number of trees
- atmospheric studies
- epiphytes, particularly manipulative studies which move pendants to various heights, or that measure conditions where the epiphytes are
- insects and their ecology and how birds affect populations

Figure 3 David Shaw measuring branch tip elongation.

Figure 3 Researchers (Bruce McCune, Oregon State University) studying the lichen communities of dead wood and snags. These areas are literally impossible to access with traditional climbing techniques. McCune and his associates described a new canopy epiphyte community associated with the top 1 m of the snags and tree tops that had never been described before.
Photo: Mark Creighton

- small animals such as birds and bats
- studies which require access to the outer crown envelope of trees
- pollination ecology
- studies on standing dead trees, they are very difficult to access any other way!

However it is not good for studying the following things:
- wide ranging vertebrates which have large home ranges
- studies which require replication in a number of forest stands. However, hypotheses can be generated from these types of N=1 studies that are then taken to a larger scale and tested
- studies which require extensive destructive sampling
- heavy use of the site can be problematic for wildlife studies, i.e. bird communities may eventually be impacted by human presence, especially on the forest floor
- it is difficult to access the inner tree crowns with a gondola, so if the project requires lots of sampling of the main stem of the tree, it may not be suitable to use the crane.

We have an information systems specialist who manages databases, our website, and interacts with researchers to aid in their sharing of data. Some researchers also give the database manager their datasets to keep at the site.

In terms of international co-operation we were original co-operators with the International Canopy Crane Network (ICCN). This work is still work ongoing and so the protocols that were proposed are still being adapted. We are also members of the Organisation of Biological Field Stations (USA, but they have an international branch), and we are an AMERIFLUX site, tied into flux networks worldwide. This program is managed by the Western Regional Center of the National Institute for Global Environmental Change. The Research Manager is also a board member of the International Canopy Network, and on the Steering Committee of the Global Canopy Programme. The Co-Principle Investigator on the project (F. Meinzer) has an ongoing research program at the Panama crane site.

Table 2 Costs

EXPENSE – ONE OFF	ESTIMATED COSTS IN US$	TIME TAKEN TO BUILD
Crane purchase	660,000	
Crane installation including environmental planning and university administration	340,000	6 months to prepare site 1 week to erect crane
TOTAL	1,000,000	

On top of the one off expenses there are also annual maintenance and running costs. These are comprised of staff costs: there are 5 full-time staff at this facility including a Research Manager; Research Co-ordinator; Program Co-ordinator; Tower Crane Operator and Research Assistant. There is also ongoing smaller maintenance work and a large annual service which is conducted in conjunction with the Morrow Crane Company. In total the maintenance takes about 200 hours a year.

Funding

Core funding has been obtained from governmental sources. The US Forest Service, Pacific Northwest Research Station provides a grant to the University of Washington for $500,000 a year. The University of Washington provides additional support.

The crane administration itself does not fund projects except perhaps in the form of waived charges for the use of crane time. A charge for crane time of $182/hr is requested, in practice however it is not always charged, and is often negotiated to an amount payable by the researcher.

Contact details

Wind River Canopy Crane Research Facility, College of Forest Resources, University of Washington, 1262 Hemlock Road, Carson, Washington 98610 USA. Phone 509-427-7028. Fax 509-427-7037. Email: dshaw@u.washington.edu.

Director of Facility: Jerry Franklin, College of Forest Resources, University of Washington Box 352100, Seattle, Washington 98195-2100 USA. Phone: 206-543-2138. Fax: 206-685-0790. Email: jff@u.washington.edu.

Co-principle Investigator: Frederick Meinzer, USDA Forest Service, Pacific Northwest Research Station, Forest Science Lab, 3200 SW Jefferson Way, Corvallis, Oregon 97331 USA. Phone: 541-758-7798. Fax: 541-758-7760. Email: fmeinzer@fs.fed.us. Web site: http://depts.washington.edu/wrccrf

Tropical Canopy Cranes

Cairns Canopy Crane, Australia

By
Nigel E Stork

Location:
'Daintree'
rainforest, North
Queensland,
Australia.
Length of
operation:
3–4 years.
Grid reference:
16°17'S
145°29'E

THE use of industrial cranes in tropical forests has opened up the canopy to exploration by scientists in the same way that the deep-sea submersible gave access to the sea floor. With the installation of a canopy crane in the 'Daintree' rainforest, Australia is now part of an international canopy crane network that is providing new information on a previously unknown part of our forests. This crane site is unusual because we have built a first class field station alongside the crane and this offers six air-conditioned bedrooms, labs, and kitchen on site for researchers who want to use the crane – none of the other canopy cranes have these kinds of facilities.

Choosing a location

The criteria which dictated the choice of location were:
- high biodiversity and conservation values since any data that might result from the crane research needed to be relevant to local and national needs
- a flat area that was unlikely to flood – a sloping site would make installation difficult and would mean the crane would need to be taller
- access to other facilities such as accommodation and tourism ventures for a possible interpretive centre
- an area that had not been logged in recent times was needed as most lowland forest in North Queensland has been logged in the last 100 years.

About a dozen sites were looked at both within the World Heritage Area and outside. An offer to lease a large area of forest for the crane facilities at almost zero cost was made by the owners of a rainforest tourism resort, Coconut Beach Resort, and this site met the criteria above.

The canopy crane is located in forest abutting the Daintree National Park at an altitude of 40 metres and less than two kilometres south of Cape Tribulation, famous for being where Captain Cook's ship "Discovery" ran aground on the Great Barrier Reef. This area is home to many of the rare and endangered species of plants for which the region is famous. The crane is located about 300 m from the forest edge and access is via a small walking track. One aspect of the canopy crane that surprises first time users is the quietness of the ride since the generator that powers the crane is several hundreds of metres away outside the rainforest. All you can hear is the noise of the insects and birds.

One reason for not trying to work within the World Heritage Area and National Parks themselves was because of the restrictions that might be placed on the operation and because government bureaucracy might slow the process of crane installation.

Even installing a canopy crane in close proximity to a World Heritage Area presents many unique problems. For example, a planning application had to be made to the local council, Douglas Shire Council, showing how the crane would meet the strict regulations with minimal environmental and visual impact. To assess the visual impact on the region large helium balloons were raised 50 m and 100 m above the site where the crane was going to be located and photographs were taken from various vantage points up to several kilometers away including photographs from out at sea. These pictures demonstrated that the balloons were not visible or barely visible and that the crane would not impact the visual amenity of the rainforest. Similarly, great efforts were made to ensure that there was minimal impact on the rainforest itself. A narrow gravel track was constructed for access, using a special fabric underlay beneath the gravel surface to spread the machine's load. One large culvert was built in the track to cover a 30 cm wide black bean vine. The route of the track was carefully planned to avoid rare and endangered species of plants in the area.

The crane itself was located in a natural gap where there had been an earlier tree fall in order to avoid unnecessary clearing. Four corner concrete pads were laid in the cleared gap and all the other crane parts, totalling 42 separate loads were carefully lowered by helicopter through the gap. We had in mind that at some point in the future the crane might need to be removed and reconstructed elsewhere and therefore the 25 tonnes of concrete at the base of the crane was lowered in more than a dozen pre-set concrete blocks of about 2–3 tonnes each.

We sought the advice of Dr Ken Chapman, the Managing Director of Skyrail, a seven kilometre cable tourist ride near Cairns, and he suggested that very few helicopter pilots around the world had the necessary skill to construct a crane. We were therefore limited to only a couple of companies in the Australasian region. The company selected, Hevilift, used a Russian Kamov helicopter with counter rotating blades to lower equipment into place. The costs of the helicopter were as expensive as the brand new crane but the skill and speed with which this crane was constructed probably shows how necessary it was to pay this amount.

One of the unfortunate features of using the helicopter was that the down force of the blades had quite an impact on one part of the canopy and scorched it quite seriously. When the helicopters lower their loads they come in at an angle to the site and it was in this area that wind damage occurred. Another critical factor was that the helicopter used a 50 m line to lower the equipment and since some parts were lowered to ground level this meant that on about twenty occasions the helicopter came within just a few metres of the top of the canopy. It was very disappointing to see the damage but all this was put into perspective when the canopy crane was hit by a Category 3 Cyclone "Rona" in February 1999 as discussed below!

Forest description

The Australian canopy crane, the first to be erected in the southern hemisphere, was installed into the Wet Tropics of Queensland World Heritage Area. Although today rainforest occupies only about 0.2% of Australia's land mass, some 30

million years ago about one third of the continent was covered. These rainforests supported a highly diverse marsupial fauna including flesh-eating kangaroos. As the continent moved northward and became drier the rainforests shrank. Indeed, during the last glacial maxima, these rainforests shrank even further than their current extent. Today about half of Australia's rainforests are in a tropical belt 400 km long and 2 to 50 km wide where the mountainous Great Dividing Range meets the sea, and the forests meet the World Heritage listed Great Barrier Reef. These forests support about 800 species of vertebrates, many of which are endemic, and about 4–5 thousand species of plants at least 1,700 species of which are endemic. They are also of evolutionary significance as 12 of the world's 19 families of primitive flowering plants are found there, two of which are endemic. This compares with the Amazon basin where just nine families of primitive flowering plants occur, none of them endemic. This is one of the few areas in the world where the reef meets the rainforest and the only place where two such World Heritage areas sit side by side.

There is a strong wet season with most rain occurring from December to April, although rain occurs in all months. The average rainfall for the site is 3.5 m a year but in 2000 a total of 6.8 m fell. Northern Australia is subject to cyclones in the wet season and their occurrence is unpredictable. On February 14 1999, just two months after installation a category 3 cyclone (Cyclone Rona) with wind gusts of up to180 kph hit the coast about five kilometres south of the crane site. The site was severely damaged with perhaps as many as 10% of the trees being brought down and 50% of remaining trees having their tops snapped.

Prior to the cyclone canopy cover was almost complete and there were few places to bring the gondola to the ground but the cyclone opened up the canopy enormously. Even though two large trees came down right next to the crane and destroyed the equipment shed, the crane was untouched. Prior to the cyclone about 30–50% of the canopy was covered by vines including lawyer cane (*Calamus* spp.), but the cyclone brought virtually all vines to the ground making access to the crane site very difficult. However the cyclone did provide a wonderful opportunity to study the natural recovery of the rainforest and now, three years on, it would be very difficult for most people to know that there had ever been a cyclone.

A Geographical Information System (GIS) of the site has the identifications, locations, heights and sizes of all trees of trunk diameter larger than 10 cm mapped. There are about 600 trees of this size, numbering 80 species. Replication is not a problem as there are three or more individuals for half of these species.

Table 1 Crane vital statistics

Make	Liebherr 91EC
Fixed/mobile	Fixed on four concrete corner pads
Tower height	47.5 m
Jib length	55 m
Max. height reached with gondola	Circa 40 m
Gondola type	1.5 m x 1.5 m and 2.3 m high purpose built
Number of people carried	Max. 3
Area forest accessed	1.0 ha

The commonest tree with 80 individuals is the black palm (*Normanbya nomanbyi*). The heights and rough dimensions of trees have been plotted on the GIS. Similarly all epiphytes have been named and plotted. No comprehensive survey has yet been undertaken of the vertebrates at the site although a species list has been developed. This includes the Southern Cassowary, Lumholtz tree kangaroo, and feral pigs. A pair of ospreys nest on the ballast weight of the crane jib and have successfully bred there.

Research projects

In the last three years more than 20 canopy crane projects have been reviewed and approved by a research committee. The committee carefully reviews projects to ensure that they will not adversely affect the site or other projects and that they increase collaboration. Funding for research has come from a variety of sources including the Cooperative Research Centre for Tropical Rainforest Ecology and Management (the Rainforest CRC), Australian Research Council, Australian Geographic and Fuchs Oils. Some of the key projects are described below.

A major focus for many of the research projects is the discovery, mapping and identification of the organisms found in the canopy. This includes surveys of mites, insects, fungi and epiphytes. In the last year more of the projects undertaken using the Australian crane are being replicated elsewhere on other cranes.

Studies on insects are focussed on the way these organisms interact with each other and with plants, their diversity and distribution, and their role in canopy processes. I am looking at how many species of insects there are in this forest and what proportion are found in the canopy. Some researchers believe that the canopy is twice as rich in insect species as the ground but this has not yet been tested. At four locations on the canopy crane site and at one location 100 metres away, pairs of combined Malaise-Flight Interception Traps have been placed in the canopy and directly below on the ground. These traps catch flying insects, some flying up into the top part of the net where they are caught in a bottle and some dropping into a water trap at the bottom of the net. Sampling two weeks a month, these traps have caught about 10,000 beetles that are being sorted into species. Other projects include a replication of Dr Odegaard's study on the host-specificity of insects previously carried out at the Panamanian crane site, and measurements of leaf herbivory.

Professor Roger Kitching and collaborators from the University of Leipzig are studying the dynamic interactions between the flowering cycle of the principle trees and vines on the crane plot and the assemblage of canopy arthropods that occur in and around the inflorescences. Currently a complete year of data is being assembled on

Access to the Australian rainforest canopy north of Cairns. Photo Michael Cermak, Rainforest CRC.

arthropods on inflorescences with spin-off information on flowering phenology, pollen structures and casual visitation of flowers by larger insects. The assemblages of arthropods associated with flowers will be sorted into guilds and their roles as potential pollinators, nectarivores, anthophages, predators and so on, assessed. Hypotheses derived from these phenomenological studies will be further tested using exclusion experiments. Nico Blüthgen, a PhD student from the University of Bonn, is looking at the use of plant food resources by ants. He has found that 40% of the tree species have extrafloral nectaries which are used by ants. He is also examining the preferences of ant species for different sugars, particularly those of the dominant ant species on the site, *Oecophylla smaragdina*, which also tends Hemiptera. This study will help determine whether ant mosaics occur in natural tropical rainforests.

A second major focus relates to the broad issue of tropical rainforests and their role in photosynthetic processes and climate. A team of researchers (Associate Professor Turton, and Drs Franks, Liddell, and Tapper) are studying how carbon, heat and water fluxes change over time as the canopy recovers from the cyclone. They are also trying to resolve which are the main species responsible in shifting the carbon balance. This research is comprised of three components: 1) ecosystem-level CO_2 heat and water flux (eddy covariance) measurements, 2) meteorological and micrometeorological measurements, 3) leaf-level leaf canopy photosynthesis measurements, and 4) soil CO_2 flux measurements. Cyclone Rona dramatically altered the micro- and meso-climate of the area by removing almost all foliage and bringing the canopy boundary layer almost to ground level. Over the last three years, regrowth of the canopy has resulted in an increased difference between the maximum temperature at canopy and ground levels.

The canopy crane is also being used by Dr Stuart Phinn (University of Queensland) and Dr Alex Held (CSIRO) to assist the development of operational methods for monitoring of the condition of tropical rainforests. In particular the crane is being used to provide access to the canopy to test the different signals from various new remote sensing technologies. A library of canopy reflectance signatures was collected from the main species at the crane site using measurements taken from the crane. These data have been used to successfully develop and validate a unique approach for delineating individual crowns or groups of crowns from high spatial resolution airborne images, enabling maps of canopy diversity and density to be produced. As a result of the number intensive field and image data collection programs conducted at the canopy crane site it has been established as an international calibration and validation site for a number of new and soon to be launched earth monitoring satellites, including NASA's Pacrim II airborne science mission in 2000 and the Earth-Observing 1 system. Fieldwork conducted at the site from the crane will be used to validate approaches taken to processing images from these new satellites in rainforests. Another advantage of the critical mass of field, airborne and satellite image data for this region is the development and assessment of new techniques for monitoring tropical forest environments that will be used by the Wet Tropics Management Authority and other tropical forest monitoring agencies around the world.

Costs

The total cost was about A$1.6m. Both helicopter time, at about A$150,000 and legal and planning costs associated with getting permission to install the crane, at about A$100,000, were both much higher than had been originally thought.

Funding

The purchase and construction of the Australian crane and associated laboratory and accommodation facilities was funded through an Australian Research Council infrastructure grant with matching funds from a consortium of three universities: James Cook University, Griffith University and the University of Queensland.

Contact details

Prof. Nigel E. Stork, Chief Executive Officer, Australian Canopy Crane Pty Ltd, and James Cook University, Cairns Campus, PO Box 6811, Cairns Qld 4870
Email: Nigel.stork @jcu.edu.au, Tel 0740 421246, Fax 0740 421247

Fort Sherman and Parque Metropolitano Canopy Cranes, Panama

By
S. Joseph Wright

Location
A. Parque Natural
Metropolitano,
Panama
B. San Lorenzo
Protected Area,
Panama
Length of
operation:
A. 12 years
B. 5 years.
Grid reference:
A:
8°59′N, 79°55′W
B:
9°17′N, 79°55′W

THE Smithsonian Tropical Research Institute (STRI) initiated the Tropical Forest Canopy Program (TFCP) when a powerful new method of canopy access using construction tower cranes promised answers to several of the most vexing problems in tropical biology. Most leaves, flowers, fruits, and associated animals are inaccessible in tall forests because they are located high above the ground on the distal tips of small branches. In situ study of the physiological controls of primary production and interactions between herbivores and leaves, pollinators and flowers, and seeds and their predators and dispersers, which occur tens of meters above the ground, had been virtually impossible.

A construction tower crane erected in a forest overcame this problem by raising a gondola holding biologists and their equipment high above the forest then lowering them to a precise location in the canopy. The tips of small distal branches, which were previously inaccessible, became the easiest part of the forest to reach. Small gondolas, which were originally designed to enter chimneys, permit access to lower levels within the canopy. For the first time it was possible to make *in situ* measurements of the metabolism of canopy leaves and the behaviour of canopy insects with ecologically meaningful sample sizes.

Choosing a location

The two cranes in Panama are located in contrasting dry and wet tropical forests. The first crane was erected in dry tropical forest with a canopy height of 20–35 m, in the Parque Natural Metropolitano within sight of the Pacific Ocean. Annual rainfall averages 1,740 mm, with a strong five-month dry season from December into May. The second crane was erected in wet tropical forest with a canopy height of 30–45 m, in the San Lorenzo Protected Area within sight of the Caribbean Sea. Annual rainfall averages 3,200 mm, with a mild three-month dry season from January through March. Moisture availability controls the distributions of many tropical organisms; and, even though the two cranes are just 70 km apart, there is not a single tree species shared between the two sites. The identities, locations, and sizes of all trees found beneath the cranes can be found at http://canopy.stri.si.edu.

Table 1 Crane vital statistics (A: Parque Metropolitano; B: Fort Sherman)

Make	A. Potain B. Kroell
Fixed/mobile	Fixed
Tower height	A. 42 m B. 52 m
Jib length	A. 52 m B. 54 m
Max. height reached with gondola	A. 39.5 m B. 49.5 m
Gondola type	Lifting Technologies © Man Baskets
Number of people carried	1, 4, or 6
Area forest accessed	A. 0.85 ha B. 0.92 ha

Research projects

The TFCP supports a wide diversity of research. Major research efforts have focused on biodiversity, trace gas evolution, interpretation of remotely sensed signals, and the physiological controls of plant function. The accumulated data makes ever more refined research possible. Data already available include many thousands of records of insect-plant associations and reproductive phenologies, leaf lifetimes, photosynthetic potentials, and seed masses for virtually all tree and liana species. Publications resulting from the program can be found at http://canopy.stri.si.edu. A handful of significant discoveries are highlighted here.

Refining measures of total global species diversity

It is widely recognized that the number of species inhabiting our planet is not known to the nearest order of magnitude. Scientists have described just 1.8 million species, while estimates of the number that remain to be described range from an additional 2 to 100 million species. Much of the uncertainty concerns insects, mites, and spiders that inhabit tropical forests and particularly the canopies of tropical forests. Entomologists with the TFCP have documented vertical stratification of insects between the herb, shrub, and tree layers within the forest and thus refined estimates of the specificity of herbivorous insects for particular species of host plants. Vertical stratification was much greater than anticipated with a completely distinct assemblage of herbivorous insect species found feeding on understory saplings and canopy trees of the same tree species(Basset 2001a). Host specificity was, however, much lower than anticipated (Ødegaard et al 2000). The largest estimates of the global number of species are extrapolations, which multiplied the known number of plant species by estimated levels of insect host specificity. The newly refined estimates of host specificity greatly reduced the estimate of global number of species and contributed to an emerging consensus that from 2 to 5 million species of insects, mites, and spiders remain to be described in tropical forests (Basset 2001b, Ødegaard 2000).

Parque Metropolitano crane, in dry tropical forest, Panama. Photo: Andrew Mitchell.

Forests as a source or sink for atmospheric gases

Forests absorb and release a wide range of gases with profound effects on atmospheric chemistry and global energy cycles and with implications for global climate change. Green plants absorb carbon dioxide during photosynthesis and release carbon dioxide during respiration and when they decompose. The atmospheric concentration of carbon dioxide has risen inexorably since the beginning of the Industrial Revolution and is expected to double in the next century with the burning of fossil fuels. If forests absorbed more carbon dioxide than they released, then forests would

mitigate these anthropogenic releases of carbon dioxide. To evaluate this possibility, global change scientists with the Tropical Forest Canopy Program doubled atmospheric concentrations of carbon dioxide for branches of canopy trees. Branches fertilized with carbon dioxide quickly failed to export the products of elevated photosynthesis to other parts of the tree, photosynthetic enzymes were then down regulated, and the amount of carbon dioxide absorbed quickly fell back to levels observed in control branches, which were not fertilized with carbon dioxide (Lovelock *et al* 1999). This elegant experiment demonstrates that tropical forest trees will not mitigate anthropogenic releases of carbon dioxide in the next century.

Forests also absorb and release smaller amounts of other gases including isoprene, a volatile hydrocarbon, and nitrous oxides, a family of potent greenhouse gases. Isoprene is released by leaves and quickly decomposes in the atmosphere to form hydroxyl ions, which control the reduction status of the atmosphere. Plant physiologists with the TFCP discovered that tropical trees emit unexpectedly large amounts of isoprene and absorb unexpectedly large amounts of nitrous oxides. These discoveries will lead to significant improvements in models of the global atmosphere (Lerdau & Throop 1999, Sparks *et al* 2001).

How do plants raise water to the high canopy

One of the great puzzles of the natural world is how tall trees and lianas raise water to leaves 30 to 50 m above their roots. In one of the first applications of canopy cranes, plant physiologists inserted micro-pressure transducers into individual xylem elements of canopy leaves to evaluate the theory that water ascends under negative pressures, maintained by evaporation at the stomata (Zimmermann *et al* 1994). This research stimulated a complete evaluation of the ascent of water in tall trees, which documented an unexpected role for water storage in xylem elements at intermediate heights during the night and which supported the theory of negative xylem pressures (Meinzer *et al* 1997).

Optimal leaf lifetimes

Canopy plants face a wide array of physiological stresses. Plant physiologists working with the TFCP have shown that young leaves can be irreversibly damaged by the high light intensities observed on clear days in the tropics (Krause *et al* 1995) and that several physiological mechanisms protect older leaves from similar damage (Veit *et al* 1996). Optimal leaf lifetimes may represent a trade off between the carbon gained through photosynthesis and the carbon costs of leaf construction and maintenance, which is mediated by the senescence of leaf function. Plant physiologists working with the TFCP have provided the strongest support for this theory (Kitajima *et al* 1997). The theory is also being extended to incorporate changes in the light environment that occur as newly produced leaves cast shade on older leaves, which further reduces the opportunity for photosynthesis in older leaves (Kitajima *et al*, 2002).

Additional possibilities for future research are virtually unlimited. The opportunity exists to conduct *in situ* studies of pre-dispersal seed predators, pollinators, herbivores that feed on pollen and flowers, parisitoids, sap-suckers, and wood-borers; to conduct feeding experiments in the canopy and understory; and to evaluate the role of ants in structuring arboreal communities, the seasonality of insect herbivores and their host plants, and the importance of ephemeral and long-lasting habitats for arboreal arthropods. The two cranes offer the additional opportunity to conduct comparisons among the canopies and understories of contrasting wet and dry tropical forests.

Funding

STRI and the United Nations Environment Program (UNEP) collaborated to erect both canopy cranes in the Republic of Panama. UNEP joined the Tropical Forest Canopy Program to provide technical support to the International Convention on Biological Diversity and the United Nations Framework Convention on Climate Change. The governments of Belgium, Denmark, Finland, Germany, and Norway contributed funds through UNEP to purchase two cranes and to administer the program for six years. STRI provided funds to erect the cranes and continues to administer the program. STRI facilities are open to scientists from all nations. Descriptions of STRI and the Tropical Forest Canopy Program can be found on the worldwide web at www.stri.org and http://canopy.stri.si.edu, respectively.

The Smithsonian Tropical Research Institute encourages scientists from all nations to take advantage of these research opportunities. The Institute provides visiting scientists with housing, laboratories, and selected research equipment and obtains appropriate research permits from the Government of the Republic of Panama. STRI awards grants-in-aid of research to support selected students, scientists from developing countries, and initial exploratory trips. A wide range of fellowship programs also support pre- and post-doctoral investigators. These opportunities are described on the worldwide web at www.stri.org.

Using a crane, individual leaves on the outermost edge of the canopy can be reached repeatedly and with bulky equipment, carried in the crane gondola. Photo: Andrew Mitchell.

Contact details

S. J. Wright, Tropical Forest Canopy Program, Smithsonian Tropical Research Inst, Apartado 2072 , Balboa, Ancon, Republic of Panama

S. J. Wright, Tropical Forest Canopy Program, Smithsonian Tropical Research Inst, Unit 0948, APO AA 34002-0948, USA

Email: canopy@stri.tivoli.edu

Literature cited

Basset Y (2001)a. Communities of insect herbivores foraging on saplings versus mature trees of *Pourouma bicolor* (Cecropiaceae) in Panama. *Oecologia* 129: 253–260.

Basset Y (2001)b. Invertebrates in the Canopy of Tropical Rain Forests. How much do we really know? *Plant Ecology* 153: 87–107.

Kitajima K, Mulkey SS & Wright SJ. (1997). Decline of photosynthetic capacity with leaf age in relation to leaf longevities for five tropical canopy tree species. *American Journal of Botany* 84: 702–708.

Kitajima K, Mulkey SS, Samaniego M & Wright SJ. (2002). Decline of photosynthetic capacity with leaf age and position in two tropical pioneer tree species. American Journal of Botany in press.

Krause GH, Virgo A & Winter K (1995). High susceptibility to photoinhibition of young leaves of tropical forest trees. *Planta* 197: 583–591.

Lerdau M & Throop H (1999). Isoprene emission and photosynthesis in a tropical forest canopy: Implications for model development. *Ecological Applications* 9: 1109–1117

Lovelock CE, Popp M, Virgo A & Winter K (1999). Effects of elevated CO_2 concentrations on photosynthesis, growth and reproduction of branches of the tropical canopy tree species, *Luehea seemannii* (Tr. & Planch.). *Plant, Cell and Environment* 22: 49–59

Meinzer FC, Andrade JL, Goldstein G, Holbrook NM, Cavalier J & Jackson P (1997). Control of transpiration from the upper canopy of a tropical forest: the role of stomatal, boundary layer and hydraulic architecture components. *Plant, Cell and Environment* 20: 1242–1252.

Ødegaard F, Diserud OH, Engen S & Aagaard K (2000). The magnitude of local host specificity for phytophagous insects and its implications for estimates of global species richness. *Conservation Biology* 14: 1182–1186.

Ødegaard F (2000). How many species of arthropods? Erwin's estimate revised. *Biol. J. Linn. Soc.* 71: 583–597.

Sparks JP, Monson RK, Sparks KL & Lerdau M (2001). Leaf uptake of nitrogen dioxide (NO_2) in a tropical wet forest: Implications for tropospheric chemistry. *Oecologia* 127: 214–221.

Veit M, Bilger W, Mühlbauer T, Brummet W & Winter K (1996). Diurnal changes in flavonoids. *Journal of Plant Physiology* 148: 478–482.

Zimmermann U, Meinzer FC, Benkert R, Zhu JJ, Schneider H, Goldstein G, Kuchenbrod E & Haase A (1994). Xylem water transport: is the available evidence consistent with the cohesion theory? *Plant, Cell and Environment.* 17: 1169–1181.

Lambir Hills Canopy Crane, Malaysia

By
Shoko Sakai,
Tohru
Nakashizuka,
Tomoaki Ichie,
Masahiro Nomura
and Lucy Chong

Location: Lambir
Hills National Park,
Sarawak, Malaysia
Length of
operation: 2 years.
Grid reference:
4°2'N, 114°55'E

THERE are two principal canopy access facilities at the Lambir Hills NP. The original one is a tower and walkway system built in 1992 with financial aid from the Ministry of Education, Science, Sports and Culture, Japan. This system consists of two tree towers connected by nine lengths of aerial walkway totalling about 300 m (Inoue *et al* 1995). Additional aluminum ladders and terraces on most post trees enable access to the canopies of trees which are up to 70 m above the ground making them 30–50 m higher than the label walkway (20–40 m high). The other facility is the canopy crane (described below) constructed to enable a greater variety of research to be undertaken including three-dimensional monitoring of the physical environment.

Choosing a location

After a general survey of topography and forest structure, the following criteria that dictated the choice of location were:
• need for a primary forest location
• flat enough area for the construction of the crane
• need for a reliable supply of electricity
• obtaining permission from the government

In total four potential locations were looked at. The first of these did not receive approval from the Sarawak government and so we decided to erect the crane at the second location. This site is about 20 minutes' walk from the lab at the edge of the forest.

We selected a construction company Konoike Construction of Japan who had the expertise and ability to build the crane and who offered us a reasonable price. After consultation with them and after taking into account the planned research, cost of the equipment and the height of the trees, it was decided to use a crane from the Liebherr company in Germany. This was then built on the site with the aid of a helicopter (Figure 1).

Forest description

Studies on forest canopy in Lambir Hills NP started in 1992. The park is located c.30 km south of Miri, NE Borneo. The altitudinal range in the park is from 30 m at the lowest point up to the top of the highest peak which is Bt. Lambir at 458 m in altitude. Vegetation is dominated by lowland mixed-dipterocarp forests on the hills with a small area surrounding Bt. Lambir covered by kerangas (heath forest) vegetation.

Crane vital statistics

The entire process of planning the crane was proceeded by close collaboration between Japanese researchers and the Forest Department Sarawak (FDS). The project was approved by the state government, who also granted tax exemption for the import of the crane.

Table 1 Crane vital statistics

Make	Liebherr
Fixed/mobile	Fixed on a concrete anchor and base
Tower height	85 m
Jib length	75 m
Max. height reached with gondola	Circa 75 m
Gondola type	A. cylindrical, 2.4 m x 0.5m; B. 2.4 m x 1.2 m
Number of people carried	A. max. 1 person; B. max. 5 persons
Area forest accessed	1.77 ha

The whole construction process took about three months. The first stage was the construction of an anchor and base at the site. The main parts of the crane were installed by helicopter after their pre-assembly outside the forest. Finally, the elevator was attached to the crane tower.

Unfortunately, in this case the construction process damaged the forest around the crane. It is essential to inspect the construction, since local constructors sometimes do not understand the importance of not destroying the forest.

Research projects

The following baseline information has been collected:
- identification of plants mostly to species level, leaf morphology, and carbon and nitrogen concentration in each species
- climate data such as temperature, humidity, and wind velocity
- insect species identification
- vegetative and reproductive phenology of plant species

In the 4 ha plot underneath the crane, all the trees greater than 10 cm dbh are tagged and identified to species. Insect population dynamics are also monitored with the use of light traps. Added to this litter and seed falls are measured with traps set in the plot. Some of this information can be accessed by members of the programme through the internet. A digital information system combining topographic and vegetative data is also under construction.

Future planned research includes investigations into ecophysiological measurements (photosynthesis, photoinhibition damage, water potential and resource dynamics); plant-animal interactions (pollination and predation herbivory); canopy structure measurements and monitoring of the physical environment (solar radiation, photosynthesis, transpiration, respiration, convection and turbulence), and fluxes of carbon and water (Itioka *et al* 2002).

Costs

The total cost for the crane and construction was about $2,300,000. Maintenance of the crane requires 30 working days per year from two experts, which costs approximately $50,000. Bimonthly inspection also has to be conducted by a local company, and experts from the supplier (Liebherr Co. Germany) check the crane twice a year. Besides these costs there are also the expenses of fuel, oil and safety tools which also have to be checked daily by crane users and operators.

Funding

The Japan Science and Technology Corporation (JST) financially supported the whole programme including construction of the crane and much of the project research being conducted using it. Thus, all the research projects using the crane are under the MOU between the Forest Department and JST (funding agency) on the research program. Each project has to be approved by the Canopy Crane Steering Committee.

Figure 1 End section of Lambir Hills crane jib being lowered into position in Sarawak, Malaysia. This requires a high standard of precision flying. Photo: Tohru Nakashizuka

Contact details

Author for correspondence Tohru Nakashizuka

Research Institute for Humanity and Nature, Takashimacho 335, Marutamachi-Dori Kawaramachi, Kamikyo-ku, Kyoto, 602-0878 Japan

Tel: 075-229-6111, Fax: 075-229-6150

Shoko Sakai – Graduate School of Human and Environmental Studies, Kyoto University, Sakyo, Kyoto 606-8501, Japan, Tel: +81-75-753-6853, Fax: +81-75-753-2999, Email: sakai@bio.h.kyoto-u.ac.jp

Tomoaki Ichie – Hokkaido University Forests, Kita-9, Nishi-9, Kita-ku, Sapporo 060-0809, Japan.

Masahiro Nomura – Center for Ecological Research, Kyoto University, Kamitanakami Hirano-cho, Otsu 520-2113, Japan and CREST, Japan Science and Technology Corporation, Kawaguchi 332-0012, Japan.

Lucy Chong – Forest Department Sarawak, 93660 Kuching, Sarawak, Malaysia.

Literature cited

Itioka T, Nakashizuka T & Chong L (2002). *Proceedings of the International Symposium "Canopy Processes and Ecological Roles of Tropical Rain Forest"*. Center for Ecological Research, Kyoto University, Japan

Inoue T, Yumoto T, Hamid AA, Lee HS & Ogino K (1995). Construction of a canopy observation system in a tropical rainforest of Sarawak. *Selbyana* 16: 24–35.

Surumoni Canopy Crane, Venezuela

By Hans Winkler
and Christian
Listabarth

Location:
Surumoni River,
Venezuela
Length of
operation: 5 years.
Grid reference:
3°10'N, 64°40'W

INSPIRED by the prototype canopy-crane "Panama I" and persuaded by national researchers, the Austrian Academy of Sciences (AAS) decided to install and maintain a canopy crane in order to give tropical ecologists the chance to study the canopy of a lowland rain forest. This was the third functional canopy research crane to be erected (after Panama I and Windriver). Upon the crane's installation, the AAS donated it to the Venezuelan Ministry of environmental affairs (MARN).

Choosing a location

The criteria which dictated the choice of location were:
- pristine tropical forest which was accessible
- suitable topography to allow for crane construction
- lack of seasonal inundation for safe crane operation
- a site in the Amazonian bioregion

At the start of the project in 1993, Austrian scientists involved in tropical biology were invited to express research interests and geographical preferences for the location of a new canopy research facility.

In this light the proposed potential locations for the crane, including Costa Rica, French Guiana, Ecuador and Venezuela, were discussed. A committee of prospective project contributors finally decided on the Venezuelan site, on scientific but also political and logistic grounds.

The location was thought to be representative of a vast area north of the Amazon that differs significantly from the much studied Central American tropics. A biological research station, the "Centro Amazonico de Investigaciones Ambientales Alejander Humboldt" (CAIAAH), had just been set up in La Esmeralda by the German overseas development agency (GTZ) and was offered by the Venezuelan government to the project.

The final selection of a specific location involved visiting 'in the company of local people' several locations proposed by them. The final site chosen at the Surumoni River, a tributary of the Orinoco, constitutes the only crane site without road access and thus lies in a huge area of intact forest (Figure 1). It is located about 15 km west of the nearest village, La Esmeralda, with about 200 inhabitants.

Figure 1 Surumoni forest with the crane reaching 40 m across it. The crane moves along a 120 m rail track. This crane is currently no longer operational. Photo: Project Surumoni

In Venezuela, a local construction company was contracted to ship the crane from Caracas to La Esmeralda. The national air force generously supported this operation. The contractor also built the huts near the crane site, arranged for the power plant, and laid the tracks for the crane.

Several organizational steps were necessary to establish the project. These were: firstly to create institutional prerequisites for the project within the AAS (locate and use available personnel resources for planning, implementation, administration and accounting, define areas of responsibilities), and in consequence to formally establish cooperation with the Venezuelan government agencies (MARN, CONICIT), to involve the local community in La Esmeralda, and to seek out official and actual collaboration with the Autonomous Service for the Development of Amazonas (SADA) as the legitimate local authority and in its function as logistic counterpart.

Table 1 Crane vital statistics

Make	Liebherr EC 50
Fixed/mobile	Mobile on a 120 m rail track
Tower height	40 m
Jib length	40 m
Max. height reached with gondola	Circa 35 m
Gondola type	A. cigar-shaped, diameter 1 m; B. rectangular, wt. circa 200 kg
Number of people carried	A. max. 1 persons; B max. 3 persons
Area forest accessed	1.5 ha

The crane is powered by a 86 kV generator and is operated by the scientist with a remote control from the research-gondola. It was installed in two attempts with a helicopter from the Venezuelan air force in cooperation with Swiss experts (Figure 2). The preparation of the ground, railroads and the local infrastructure were done the Egyptian way without machines. The material was brought in boats from La Esmeralda, and the tiers were cast in a camp on the banks of the Orinoco. Though the seclusion of the site proved to be a guarantee of a healthy environment, and even though large mammals could often be observed at the site, this degree of remoteness was probably not essential from a scientific point of view and for the research that has been conducted. It made it necessary to install the crane with a helicopter, the most expensive and difficult method of erecting a crane.

Research projects

There was no preliminary data collection at the crane site until the crane had been erected except for a rough exploration of soil characteristics for technical reasons. Major projects carried out at Surumoni have investigated:
- energy and water requirements of the forest, including permanent climate monitoring and data collection of microclimatic processes
- structure, diversity and dynamics of the vascular epiphyte community
- community structure of fungi, lichens, and mosses
- establishment, function and distribution of ant gardens

- specificity of arboricolous ants
- bio-acoustics, including the assessment of the abiotic and biotic acoustic environment and adaptations in bird songs to bird height location
- ecomorphology and ecology of birds, also including the specialization of frugivores and the interactions between ambient light and plumage color

Except for a plot inventory, the mapping of plot trees, a structural assessment of the forest and phenological data of selected tree species there is little specific botanical information about the site.

There was no common database established across the research projects but working groups exchanged data, and other interesting observations freely. However, within project data banks were established to combine data (e.g. on behavior, morphology etc.). Workshops were also held to foster communication between groups although any interaction with scientists from other crane sites was purely on an individual basis. Available data are kept at the Konrad Lorenz Institute for Comparative Ethology (KLIVV), from where specific questions are routed to the project leaders.

Table 2 Costs

EXPENSE – ONE OFF	ESTIMATED COSTS IN US$	TIME TAKEN TO BUILD
Infrastructure Surumoni (all infrastructure building including expenses such as materials, transports and set up of the crane station, tracks, energy distribution)	102,000	10 months
Imported materials (generator, rails)	47,500	
Crane (incl. shipment and national transport)	450,500	
Installation of the crane (except helicopter)	194,500	
TOTAL	794,500	
ANNUAL COSTS		TIME TAKEN
Materials, spare parts and transports	4,700	
Fuels, oils and grease	5,700	
Weekly checks		2 hours
Monthly checks		1 day
Yearly inspection	9,200	2 persons – 3 days
Local personnel (crane administrator, local staff – camp administrator, boat operator, occasional help in field work and other errands)	25,700	
TOTAL	45,300	

Funding

The crane and local infrastructure were maintained, financed and run by the AAS until the termination of a bilateral agreement in August 2000. The AAS was willing to continue the cooperation and negotiate a new 5 year agreement with the intention to grant academic, administrative and financial support, given that the Venezuelan government would provide the necessary legal backing, and contribute at least partially to logistics.

However, the government was unable to guarantee reliable working conditions including work and sampling permits that were a major prerequisite for international participation in the project. At the moment all scientific activities have been stopped for legal and logistic reasons. The Venezuelan government is attempting to resume crane operation without external aid and expertise and to run the facility for the exclusive use of national scientists, with the option of future international cooperation. One of the major obstacles to be resolved in order to continue this project is a political one. The new Venezuelan constitution grants substantial rights to indigenous peoples within their homelands. However, because no provisions for the execution of these rights have been set in place, all scientific activities in Amazonas have ground to a halt.

The crane administration did not provide project funds, but funded the infrastructure and logistics. Venezuelan researchers were granted free access to the research facility (50% of the overall operable time), while non Venezuelan scientists were charged for crane-use to cover part of the operational costs.

Figure 2 Helicopter transporting jib to Surumoni site. Photo: Project Surumoni

Pros and cons of canopy cranes

At least one lesson could be learned from this experience: DO try to build the crane another way, but if you need a helicopter, DON´T employ amateurs and plan the operation thoroughly. Transport of fuel, equipment, spare parts, and not least people, remained a constant logistic problem throughout the Surumoni project. The area of forest accessible by the crane is also a crucial consideration. Generally, it seems to us that using a long jib rather than rails to achieve a sufficient area is the best idea. Using a longer jib presents fewer constraints with regard to the flatness of the terrain and ground stability. Added to this laying and maintaining the rails is expensive and the special equipment needed to keep the crane running is prone to failure. There may however be a trade-off between use of rails or a long jib with respect to the stability of the gondola, and any decision should be based on the methodological requirements for research. Finally, an oval rather than a circular shape of the plot may be more desirable in specific cases.

Cranes are excellent for assessing, mapping and monitoring of organisms and also for the study of processes that involve fixed (mostly plant) subjects/objects/resources such as flowers, fruits or leaves and the animals they interact with. A crane is particularly well suited for accessing specific canopy habitats in the periphery of tree crowns, though tree trunks are almost inaccessible. The only major disadvantage of this system is that it is fixed and thus forest access is limited to the range of the crane, which is always less than one would desire. The crane is also susceptible to closure due to high winds or thunderstorms.

Contact details

Prof. Hans Winkler, Konrad Lorenz-Institut fur Vergleichende Verhaltenforschung der Osterreichen Akademie der Wissenschaften, Savoyen Strasse 1A, A-1160, Vienna, Austria.
Tel: +431 486 2121, Fax: +431 486 212128
Email: h.winkler@klivv.oeaw.ac.at, c.listabarth@klivv.oeaw.ac.at

Literature cited

(for a complete and steadily updated list see www.oeaw.ac.at/klivv/surumoni)

Anhuf D, Motzer T, Rollenbeck R, Schröder B & Szarzynski J (1999). Water budget of the Surumoni-crane-site (Venezuela). *Selbyana* 20: 179–185

Anhuf D & Winkler H (1999). Geographical and ecological settings of the Surumoni-Crane-Project (Upper Orinoco, Estado Amazonas, Venezuela). *Anz. Math. Nat. Kl. Abt.* I. 135: 3–23

Blüthgen N, Verhaagh M, Goitia W, Jaffe K, Morawetz W & Barhlott W (2000). How plants shape the ant community in the Amazonian rainforest canopy: the role of extrafloral nectaries and homopteran honeydew. *Oecologia* 125: 229–240

Cedeño A, Merida T & Zegarra J (1999). Ant gardens of Surumoni, Venezuela. *Selbyana* 20: 125–132

Nemeth E, Winkler H & Dabelsteen T (2001). Differential degradation of antbird songs in a Neotropical rainforest: adaptation to bird height location. *Journal of the Acoustical Society of America* 110: 3263–3274.

Nieder J, Engwald S, Klawun M & Barthlott W (2000). Spatial distribution of vascular epiphytes in a lowland Amazonian rainforest (Surumoni crane plot) in southern Venezuela. *Biotropica* 32: 385–396

Schaefer MM, Schmidt V & Wesenberg J (in press). Vertical stratification and caloric content of the standing fruit crop in a tropical lowland forest. *Biotropica*

Szarzynski J & Anhuf D (2001). Micrometeorological conditions and canopy energy exchanges of a neotropical rainforest (Surumoni-Crane-Project, Venezuela). *Plant Ecology* 153: 231–239

Winkler H & Preleuthner M (2001). Behaviour and ecology of birds in tropical rain forest canopies. *Plant Ecology* 153: 193–202.

Site Safety Plans for Canopy Cranes

Information
compiled from
safety protocols
recommneded by
the Wind River and
Cairns Cranes

WE recommend that any canopy access programme defines a safety protocol which all personnel should use when working in the canopy. The purpose of the Canopy Crane Site Safety Plan is to identify hazards that will be, or could be, encountered in the operation of a crane research facility, and to identify the individuals responsible for safe operation of the facility. Also to set protocols, which when followed, will prevent accidents and promote safety. Here we offer a general set of guidelines that can be applied to all tower cranes which are set in the middle of forests for the purposes of investigating the ecology of the forest canopy, where researchers are lifted up above the forest and lowered to desired positions within and above the forest canopy in a suspended personnel basket. The Canopy Crane Site Safety Plan includes considerations for the lifting of researchers, in addition to general considerations for safe operation of a tower crane. It must however be recognised that these protocols can not cover every potential safety hazard associated with working in the areas around or on a crane. These recommendations are based on experience from the Australian and Wind River canopy access cranes.

Insurance
All operators of canopy cranes need to ensure that they:
- abide by all the relevant national legislation that pertains to the operation of such facilities and equipment
- are insured against accidents
- have public liability insurance
- get all users of the facility to sign legal disclaimers

Employee safety protocol training
- basic safety training for employees at the facility is essential, especially for seasonal employees
- standard First Aid and CPR training should be given
- familiarize all users with the safety plan
- familiarize all users with the rescue technique from the gondola, equipment, and where it is stored. Full training in rescue is reserved for regular users or the principal safety officer
- familiarize all users with facility layout, gates, and access routes
- explain emergency procedures, medical, fire or structural.

Areas of hazard potential when operating tower cranes
Potential problems in operation of the crane include; structural failure, motor failure, contacting jib with obstacles, failure of cable, dropping suspended personnel basket, and failure of slewing bearing. The following inspections, test lifts, maintenance, operation guidelines, and training should go a long way to prevent any problems from emerging.

Inspections

Inspections should be made daily by the crane operator following the guidelines laid down by the manufacturer of the crane and local Health and Safety Dept. protocols. Checks should be made of bolts, welds, rust, structures, motors etc. before beginning operation. After turning the crane 'on' the operator will make sure all operations seem to be running smoothly. The tower crane will also be inspected by manufacturers representatives; after erection, 30 days after erection and then as needed.

Personnel basket hoisting protocol/checklists

Prior to hoisting personnel a series of safety checks are performed. The purpose of this is to be sure the crane is still capable of supporting the specified weight for the personnel basket, all systems are still safe, all users have been instructed and that no obstacles exist in general movement of the crane. Full orientation of users occurs for each group of new users throughout the day.

Maintenance

Maintenance of the tower crane is the responsibility of the crane operator and follows general guidelines of the manufacturer. Maintenance occurs on a daily, weekly, monthly and annual basis. The crane manual will list the maintenance and inspection schedule.

Operation

A daily log will be kept by the crane operator, which is kept in the operator's cab or in the office on the site. Formal safety committee meetings will discuss any perceived problems with operation of the facility.

Training

Training is directed at employee responsibilities:
- full time arbornauts/crane operator
- office manager
- part time operators
- seasonal employees.

Fall protection

Fall protection is paramount. Each individual entering the suspended personnel basket should have a helmet, full body harness and safety rope. The safety rope will be secured to specified attachment points on the gondola.

Suspended personnel basket

The site safety officer/arbornaut will provide orientation training to any individuals who will be lifted in the basket. This will include fitting of the harness, checking harness for adequate buckle lock, and reviewing safety standards and appropriate behaviour in the suspended basket.

Tower crane

No harness and rope is required on the tower portion of the crane, within the operators cab or on the counterbalance jib. A harness and rope is required on the mast, load jib, or when climbing on structure of the crane outside bounds of usual movement.

Electrical

All systems must be checked regularly in accordance with manufacturers instructions. Personnel operating the facility must know the location of control panels and switches in case of emergency. There should be a back-up power supply facility.

Overload protection switches are provided to prevent accidental or deliberate overloading of the tower crane's structure or machinery. These overload protection switches must be set to the crane manufacturer's recommended limits after the crane has been erected and load tested.

Wildfire

Wildfire is a potential problem in surrounding forests, and in the crane stand. If smoke is observed that indicates potential wildfire, evacuate the crane facility, and cut power to it. Fire tools that should be kept at base of crane include: shovel, fire extinguisher, and during the fire season, a bladder bag.

Electrical storm

In the event of a lightening storm the gondola will be set down immediately, and cleared of all personnel until weather conditions have changed/cleared enough to allow continuation of safe operation. Crane operator and arbornaut will assess weather conditions and make the decision when, and if operations will proceed.

Additional equipment

Any additional equipment to be attached to the tower or put in the basket has to be approved by crane operator and safety committee prior to loading the gondola. Potential hazardous equipment would include items such as batteries, heaters, and pressurized gases.

Night operations

Predetermine and document coordinate points during daytime, prior to night operations taking place. A person is to be at the canopy hut for emergency contact needs. Weather should be at an optimum, and during gondola movement, a light is to be on as a visual aid to assist crane operator.

Electrical equipment

Special care should be taken and pre approved by the arbornaut while using electrical equipment in gondola to avoid electrical hazards.

Single rope techniques (SRT) can be used to escape from the gondola should the crane cease to function for any reason. This requires specialist training. See page 13–23 for more on SRT. Photo: Project Surumoni

HOIST With forearm vertical, forefinger pointing up, move hand in small horizontal circle.

LOWER With forearm extended downward, forefinger pointing down, move hand in small horizontal circles.

TOWER TRAVEL Arm extended forward, hand open and slightly raised, make pushing mothion in direction of travel.

TROLLEY TRAVEL Palm up, fingers closed, thumb pointing in direction of motion, jerk hand horizontally.

STOP Arm extended, palm down, move arm back and forth.

EMERGENCY STOP Both arms extended, palms down, move arms back and forth.

SWING Arm extended point with finger in direction of swing of boom.

MOVE SLOWLY Use one hand to give any motion signal and place other hand motionless in front of hand giving the motion sugnal (hoist slowly shown as example).

DOG EVERYTHING Clasp hands in front of body.

Figure 1 Hand signals for tower crane operation

Tree climbing
Staff qualified to tree climb will help to facilitate any rescues needed.

Areas of potential hazard when working in the forest
Potential hazards when working in the forest include animals and plants; falling objects and heat. The following safety protocols should go a long way to ensuring that the risks are minimized.

Hazardous trees/branches
Large branches or trees may fall and hit people, vehicles, suspended personnel basket or the crane. Inspections should be made of the area around the crane site

prior to crane construction for any hazardous trees etc. that would impede the safe operation of the facility. When the crane is operational inspections of the trees and branches should be carried out regularly and any remedial action taken to make them safe.

Care should be taken not to work in the forest when there is high wind or threat of storms which may dislodge objects in the canopy.

Lone working

Working alone always presents increased risk. This is especially true when working in dense forest where visibility may be reduced. As a general rule working alone should be avoided, if it has to be undertaken then the following rules can be applied:
- inform someone of work location and expected time of return
- carry means of attracting attention such as whistle or radio
- ensure person has been trained in safe working practices

Flora and fauna

In many forest situations there are likely to be animals and plants that are potentially hazardous. The hazard from plants includes thorns, stings, and poisonous parts. All personnel should be trained to be able to identify hazardous plant species so that they can be avoided. Any specific treatments that are needed should be kept on site and accessible by all personnel.

The hazard from animals includes attack especially from ferals, bites from snakes, and stings from insects. All personnel should be trained what to do in the event of an encounter with one of these hazardous species; bees can be especially dangerous in the canopy. Although most snakes are harmless all species should be treated with caution and avoided.

Where possible any anti-venom should be available on site, or arrangements made for the evacuation of the casualty.

Heat exhaustion and stroke

Working conditions in the canopy can get hot and/or humid and personnel may be exposed for extended periods of time whilst in the gondola. There can be a very real risk of heat exhaustion or potentially fatal heatstroke. Precautions to be taken include wearing of hats when working in sunlight for long periods of time, carrying adequate water supplies and not working alone.

Permit zones within the Australian Canopy Crane Research Facility.

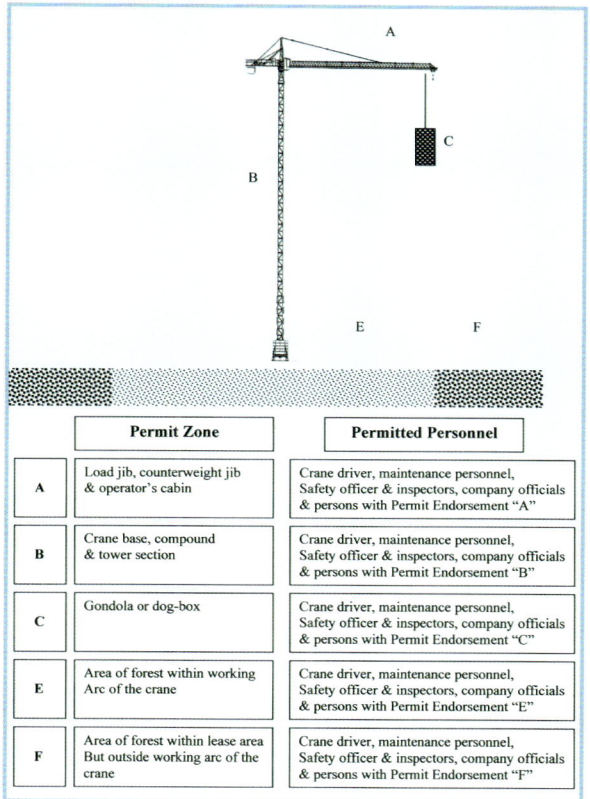

	Permit Zone	Permitted Personnel
A	Load jib, counterweight jib & operator's cabin	Crane driver, maintenance personnel, Safety officer & inspectors, company officials & persons with Permit Endorsement "A"
B	Crane base, compound & tower section	Crane driver, maintenance personnel, Safety officer & inspectors, company officials & persons with Permit Endorsement "B"
C	Gondola or dog-box	Crane driver, maintenance personnel, Safety officer & inspectors, company officials & persons with Permit Endorsement "C"
E	Area of forest within working Arc of the crane	Crane driver, maintenance personnel, Safety officer & inspectors, company officials & persons with Permit Endorsement "E"
F	Area of forest within lease area But outside working arc of the crane	Crane driver, maintenance personnel, Safety officer & inspectors, company officials & persons with Permit Endorsement "F"

Safety equipment lists for basket and operator's cab

Suspended personnel basket/operator's cab-each containing this gear

    General gear
    First aid kit
    Head lamp
    Fire extinguisher (not in gondola)
    Binoculars
    Batteries
    Bendix-King radio (emergency only, extra batteries)
    Motorola radio, SP10

Each individual:
    Helmet, body harness, safety rope

Rescue/Evacuation Gear:
    Two 300 ft rope/lifelines
    Two rope bags
    One pair leather gloves
    Eight carabiners
    Rappel rack
    Mechanical ascenders
    Webbing anchors
    Rescue pulley
    Webbing, 5 colors, 5 lengths
    Four pulleys
    Two 8 m lengths, 8 mm perlon rope, load releasing hitch
    "Edge softeners"
    Screamer suit (evacuation suit)
    Two sets, system prusiks, 125 cm and 75 cm lengths = pair, color coded

In dry shack at base of crane:
    Fire extinguisher, litter, litter harness, shovel.

Postscript

The Global Canopy Programme can accept no liability from the use or misuse of these safety protocols and strongly recommends that managers and users of canopy access sites develop their own safety protocols to suit their local needs.

4 Lighter Than Air

COPAS: A New Permanent System to Reach the Forest Canopy

By
P. Charles-
Dominique,
G. Gottsberger,
M. Freiberg and
A.-D. Stevens

Location: Les
Nouragues
Research Station,
French Guiana.
Length of
operation:
starting 2002.
Grid reference:
4°2'N, 52°40W

W HEN one of us (G. Gottsberger) thought about working in the canopy of tropical forests in the nineteen eighties, there were several systems available, with two of them having been established for a long time. These were the single-rope-technique (SRT), which was extensively used by M. Freiberg and construction cranes. The SRT was designed to work in the canopy, but going to the same spot over and over again inevitably causes some damage. The construction cranes overcome this problem, but their rigid structure is designed to work at construction sites, not in tropical forests. There are only limited ways to enlarge the area cranes can reach and depending on the crane type some damage to the forest is unavoidable, especially during the erection phase and around the base of the crane tower.

To overcome these problems, G. Gottsberger designed a new canopy access system called COPAS. The acronym COPAS stands for Canopy Operation Permanent Access System. Initially this was together with J. Döring from the University of Giessen (Gottsberger & Döring 1995) and later on with A.-D. Stevens & M. Freiberg from the University of Ulm and with the German companies Ballonbau Wörner in Augsburg, Stahlbau Glocker in Ehingen, H. Kiessling engineer's office in Konstanz, A. Hammerl engineer's office in Ellgau, R. Conradt Mess- und Regeltechnik in Allensbach, J. Kuder Industriedienstleistungen in Friedberg and the Luftfahrt-Bundesamt in Braunschweig.

The principle of COPAS

COPAS basically consists of three higher than canopy towers arranged in the corners of a regular triangle (please refer to Figure 1). Cables from the base of each tower are guided to the top of the towers and are connected at a junction knot point. A gondola moving vertically is connected to this junction knot. The three cables provide movement in the horizontal plane, the gondola itself in the vertical plane. The requirement of having the lowest possible impact on the forest led to the design of towers, which can be erected by hand. Such towers, however, would not have been strong enough to carry the weight of the gondola and its load. The solution to this was the introduction of a helium balloon pulling the junction knot of the three guiding cables vertically upwards.

After a long time designing, verifying and testing the components, the system is now ready to use and was erected for testing in the Botanic Garden of the

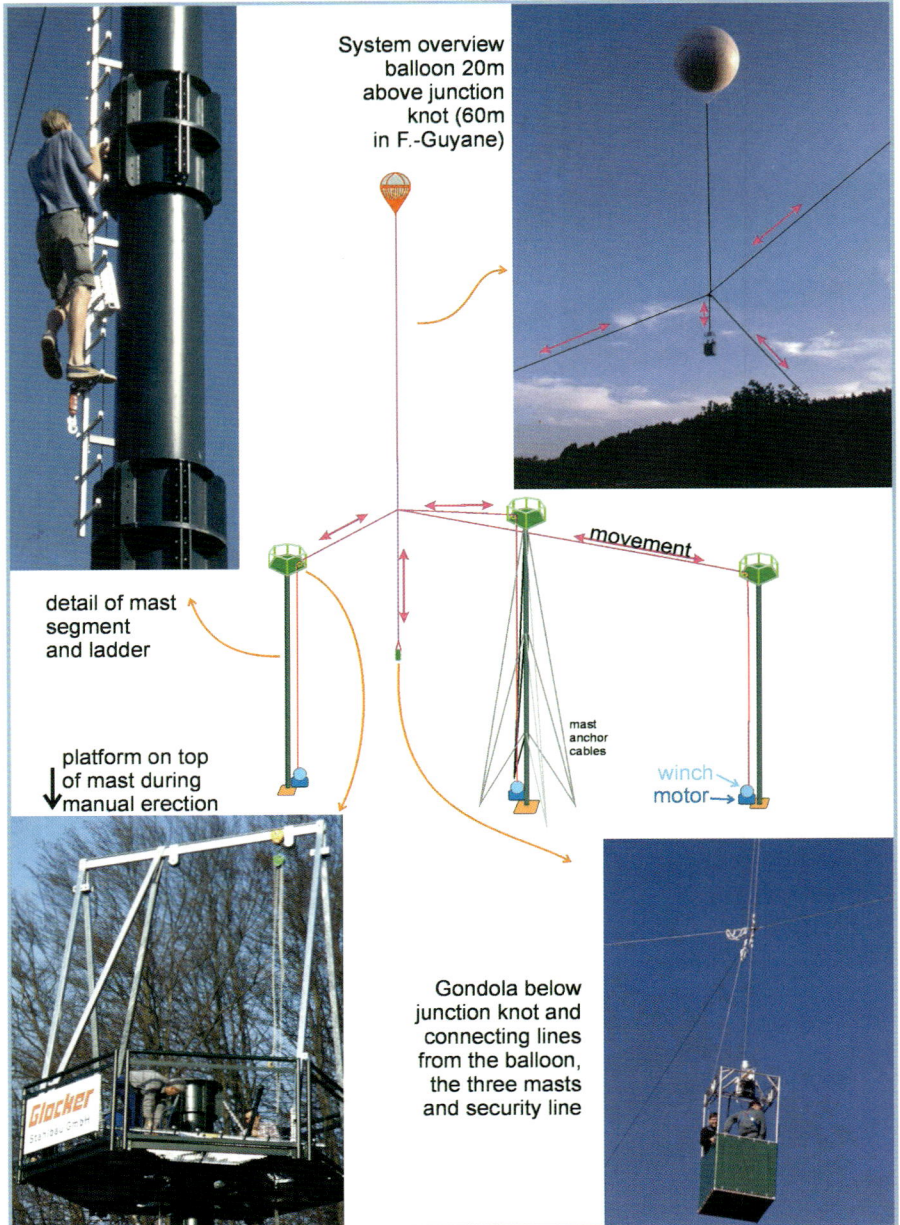

Figure 1 Flow diagram explanation of the COPAS system

University of Ulm in July 2000. The winches at the bases of the three towers have cables long enough to allow coverage of a regular triangle with a base line of 180 m. Thus the area covered is 1.4 ha. The use of a triangle makes it possible that this area can be doubled in the future by simply adding another tower. With the use of one central tower and six towers (and balloons) in the periphery, therefore, six regular triangles would cover 8.4 ha. All towers are equipped with ladders and have platforms on top, which are 3 m in diameter and which provide additional space for observations and installing various equipment. The towers are made up of 3 m long elements, mounted one above the other and secured with bolts. At 45 m the towers are a few metres higher than the surrounding forest canopy at the site in French Guiana. Each element can be installed manually and hoisted with a winch placed on the highest element already fitted. A canopy opening at the site of a tower of only 5 m diameter is enough to erect the tower.

Retracting and extending the cables of the three winches facilitates the horizontal movement of the junction knot above the canopy. If one winch is retracting its cable, at least one other winch needs to extend its cable. For safety reasons and to simplify driving commands in the gondola, the cable forces of all three winches are measured constantly by a central computer which controls their movement. The gondola pilot just pushes or pulls the gondola navigation joy stick towards the desired direction and the gondola then follows this movement. This automatic mechanism is permanently working and instantaneously corrects additional wind forces affecting the balloon or the cables.

The helium-filled balloon has a diameter of 10.8 m which corresponds to a volume of 640 m² and carries the weight of the gondola and two passengers plus equipment. The passengers are secured with harnesses at all times when on the ladders, the platforms and in the gondola. The balloon hovers 60 m above the canopy and is connected to the junction knot via cable. Even if the balloon is moving, this junction is kept in place horizontally by the computer controlled guiding cables. The balloon can be pulled to the ground by means of another winch, which is mounted on a small movable caterpillar. For climate measurements or mapping purposes etc. it is possible to disconnect the junction knot from the three mast cables and only use this winch with the balloon and the gondola and to go up 130 m above the canopy. To park the balloon at the foot of one of the towers needs a clearing 15 m in diameter, but this can be made some meters outside the investigation plot because the movable winch can be used for balloon parking. Therefore, virtually no damage needs to be done to the plot during erection or maintenance of the balloon.

Vertical movement of the gondola is facilitated by another winch within the gondola. The gondola is connected to one side of the cable of this winch, and the junction knot above the canopy is connected to the other. The electrical power for the motor is provided by a rechargable battery pack inside the gondola. Power is only needed for downward movements, and the power generated in the motor by the upward movements recharges the battery.

Time schedule

There was a long delay between the first detailed plans for COPAS in 1995 and its near realisation with installation in Ulm in 2000. The reason for this, was basically that most entrepreneurs who were used to operating in temperate regions decided that building such a device in a tropical rainforest was too risky to be seriously considered. Moreover, those that did submit tenders considerably increased their initial cost estimates. The economic recovery that occurred at the time left most firms with a heavy backlog of more substantial and conventional projects and with no wish to take any chances.

After managing some logistical, financial and bureaucratic problems in 2000, the system was transported from Germany to Cayenne harbour in French Guiana in January 2002 and will be brought to the erection site in May 2002. Erection will take place at the beginning of the next dry season, which is June/July 2002 and after a test phase of three months the whole system is expected to be fully operational by October/November 2002.

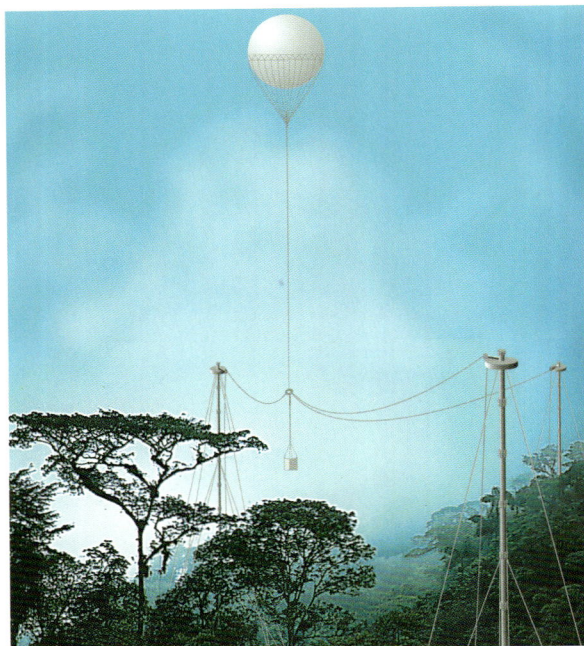

Figure 2 An artists impression of the COPAS project in operation.

Location of COPAS

For historical reasons, the legislation in the French overseas department of French Guiana, is relatively similar to that in the departments in mainland France. Therefore, it follows European law, uses the Euro, has European infrastructures including medical care and lots of other advantages. Moreover and very important is the fact that the 150 000 inhabitants mainly live along the coast and 85% of the land is covered by virgin, primary forest. This together with the excellent French-German science partnership led to the selection of French Guiana for the implementation of the first COPAS project. More precisely, COPAS will be erected in the 1000 km sq Les Nouragues Strict Nature Reserve, which is located in a completely uninhabited area, 100 km inland from the coast. The COPAS site (4°2'212 N, 52°40'654 W) was selected inside this reserve, close to the Saut Pararé rapids of the Arataye river ("Pararé-station") and 5.5 km from the Les Nouragues CNRS Research Station (see Bongers *et al* 2001).

The presence of an established research station which has been active since 1986, allows the COPAS project to be launched in a forest already relatively well studied. There were several reasons for selecting a site some kilometres away from the present research station ("Inselberg-station"). These were:

1. the relatively large diversity and abundance of epiphytes at the site compared to other sites in French Guiana

2. the site is 800 m downstream from a section of rapids, the Saut Pararé, where it will be possible to install a small water turbine to produce hydroelectric power

3. most transport between Cayenne and the Les Nouragues Research Station currently takes place by helicopter but, for large loads, river transport is generally cheaper. The development of two research sites a few kilometres apart will

lessen the human impact on both of them (2 1/2 to 3 1/2 hours are needed to walk the 7.5 km footpath).

Moreover, the two sites are complementary in terms of habitat diversity, since the Inselberg-station is located at the foot of an Inselberg and the second one (the Pararé site) lies on the banks of a river; this arrangement will therefore enable researchers to cover a wider range of biotopes.

Research projects

The Les Nouragues Research Station initially operated around a few French research projects focusing on forest regeneration and seed dispersal by fruit-eating vertebrates. Relatively recently, other topics of study have been developed, partly as a result of collaboration with scientists from other countries (mostly from Europe and North America). The web site www.cnrs.fr/nouragues provides information on the station and its activity, as well as on the COPAS project. The book "Nouragues, Dynamics and Plant-Animal Interactions in a Neotropical Rainforest" (Bongers *et al* 2001) summarizes research undertaken at the station over the last dozen years. COPAS is part of a large scale research project on the Guianian tropical rainforest ecosystem and, in conjunction with other research stations is part of the global scientific programme focussing on tropical rainforests. The new COPAS unit should allow development of ongoing research programmes and launching of new projects. The facility will be open to the international scientific community whose projects will be dependent upon approval by a selection committee, which aims to enhance collaboration with other canopy study sites worldwide. During workshops in Paris and Ulm, the following projects were selected:

- the mechanisms of tree height and diameter growth, coupling *in situ* observation with laser measurements; dynamics of tree crown growth; productivity, biomass of root-climbers and hemi-epiphytes
- investigations into sunlight characteristics; influence of the canopy on the lighting of different forest strata; consequences on the evolution of coloured signals, on biomass and on biodiversity
- microclimates and microhabitats in the canopy; water and nutrient cycles (storage in the ramosphere); the role of epiphyllous organisms in the fixation of atmospheric nitrogen
- the mechanisms of photosynthesis; correlations with microclimate and transpiration; C3 (trees) and C4 (epiphytes and stranglers) photosynthesis; the mechanisms of gas exchange within the canopy and in the underlying strata (carbon isotopic discrimination, sap flows, stomatal conductance of water vapour, balance of CO_2 and H_2O flows, ^{13}C)
- survey and ecology of a number of plant families (Annonaceae, Araceae, Arecaceae, etc.)
- pollination systems; the roles of the different insect and nectarivorous vertebrate communities; pollination syndromes; flowering and fruiting phenology
- study of predation of immature fruit; the role of arboreal rodents, parrots and seed-eating insects (biomass produced, timing of metabolite mobilisation in developing seeds, production of toxic secondary compounds)
- seed dispersal by the fruit-eating community
- ant-plant relationships (selection of nesting sites, feeding behaviour, the function

of venoms, conflicts of interest between entomophilous pollination and ant-based protective systems)
- the biodiversity of canopy invertebrates: lepidoptera and relationships with their host-plants, heteroptera, ants, termites, coprophagous coleoptera, etc.; food resources and timing of reproduction; invertebrate movements from one tree crown to another
- the sharing of resources and space between communities of arboreal vertebrates (rodents, primates, bats, sloths, birds, amphibians)
- medical entomology: investigations on the insect vectors of yellow fever, malaria, leishmaniasis and Chagas' disease; seasonal cycles of vectors and parasite infections
- survey of the reservoirs of plant trypanosomes

Funding and management

Several funding sources were used and the final success of the project was due to receipt of the Körber European Science Award in 1996 by the two project coordinators G. Gottsberger, *Abteilung Systematische Botanik und Ökologie* (Universität Ulm, Germany) and U. Lüttge, *Institut für Botanik* (Technische Hochschule Darmstadt, Germany), and further P. Charles-Dominique, *Laboratoire d'Ecologie du Muséum National d'Histoire Naturelle et Station de Recherche des Nouragues* (UPS 656, CNRS, France), A. Cleef, *Hugo de Vries Laboratory* (University of Amsterdam, the Netherlands), B. Hölldobler, *Lehrstuhl Zoologie II* (Universität Würzburg, Germany) and K. Linsenmair, *Lehrstuhl für Tierökologie und Tropenbiologie* (Universität Würzburg, Germany). Other sources came from the University of Ulm and the State of Baden-Württemberg, Germany, as well as from the CNRS, the Ministry of Scientific and Technical Research and the Regional Council of French Guiana, France. A big chunk of this money inevitably had to be used for the technical development of COPAS.

The construction of COPAS is placed under the responsibility of the University of Ulm, whereas the setting up of the base camp, the transport and the on-site assembly of COPAS are under that of ECOFOR (a group of several French scientific bodies involved in forest research). The CNRS, which also manages the Inselberg-Station, will be in charge of the management and running of COPAS.

Contact details

Prof. P. Charles-Dominique, Museum National D'Histoire Naturelle, Avenue du Petit Château 4, 91800 F-Brunoy, France

Prof. G. Gottsberger & Dr M. Freiburg, Abteilung Systematische Botanik und Ökologie, Universität Ulm, Albert-Einstein-Allee 11, D-89081 Ulm, Germany

Dr A.-D. Stevens, Botanischer Garten und Botanisches Museum Berlin-Dahlem, Königin-Luise-Str. 6-8, D-14191 Berlin, Germany

Literature cited

Gottsberger G & Döring J (1995). 'COPAS'. *An innovative technology for long-term studies of tropical rain forest canopies.* Phyton (Horn, Austria) 35: 165–173.

Bongers F, Charles-Dominique P, Forget PM & Théry M (eds.) (2001). *Nouragues. Dynamics and plant-animal interactions in a Neotropical rainforest.* Kluwer Academic Publishers, Dordrecht, 421 pp

The Canopy Raft

By
Francis Hallé

THE Radeau des Cimes (the Raft of the Tree-tops) is a general term for several methods of studying the canopy of tropical forests through the use of hot air balloons and other peripheral equipment. The first canopy visit aided by a hot air balloon took place in 1981. The first actual placement of a raft on to the canopy was in 1985 and our first tropical forest operation took place in 1986 in French Guiana. There have been six successful operations which took place between 1986 and 2001.

Choosing a location

There were a number of criteria which influenced the choice of suitable sites, broadly speaking there were ecological, political and social considerations. The canopies we study are in the humid tropics and the sites we select need to possess forest that has been well protected and is if possible primary in nature. The local government of the area also needs to be welcoming to international scientific expeditions and must not obstruct the production and publication of research findings. The scientific community of the host country has a pivotal role: if it opposes our choice of location, the operation cannot take place; if it co-operates, the operation will be a success.

One criteria which is very specific to the choice of the site for the placing of a Radeau (raft – see explanation later on) is the need to avoid secondary forest, as the trees are too fragile. Only primary forest has sufficient solidity to support this type of equipment. Within these requirements the actual choice of country is often down to the principal sponsor of the operation who will choose a country in which they have a direct interest. For example, it was Elf that chose Cameroon in 1991 and Shell which chose Gabon in 1999 for operation of the canopy raft.

Equipment and Installation

There are three main access methods used with/for this technique, each using a slightly different combination of a hot air balloon and various other equipment. The first two methods involve suspension below an airship (Figure 1)which carries the researchers around the canopy.

- the Radeau or Raft (Figure 2 is a pretzel shaped ring of inflated rubber tubing that rests on the top of the canopy trees. A net is attached underneath the tubes and connects them so that people can walk across from one side to the other. Sampling occurs at the net edge by researchers leaning over and collecting from the tree tops within reach.
- the Luge (Figure 3) consists of six 2.5 meter long inflated tubes joined to form a hexagon, the bottom of which is covered with a net. The structure acts like a basket within which 3 people can work and be transported around the canopy.

The other method two methods differ in that they do not use the airship as a means of transportation.

Overleaf: Figure 1 Airship over forest
Photo: Laurent Pyot

- the Bulle des Cimes or the Bubble of the Tree-tops (see Figure 4) uses a helium
  filled balloon to carry a single person across the tree-tops. A guide cable one
  kilometer long is laid across the top of the trees by the airship and then attached
  to the emergent crowns. The operator can then pull themselves along the guide
  wire and survey the canopy as they go.
- the ICOS (Figure 5) is an icosahedranal (twenty sided) unit that is permanently
  fixed in the fork of a large tree, and is designed to be a pod for living in. It is
  possible for people to climb into the unit as well as to sit on the platform which
  is attached to the top of it and to make observations from both these positions.
  This is good for longer term, stationary research and for forests where there is
  too much wind to use the balloons. The ICOS can be installed with the aid of
  the airship or can be assembled at the base of the tree and then hoisted into
  position.

The various assemblies described constitute a range of flexible equipment, which
allows work to be carried out whatever the local forest's conditions. In primary
forest, one can use all the machines; in secondary forest, the Luge and the Bulle are
well adapted; and in mountane forest, the Luge would be the ideal machine.

Our hot air balloons, conventional or dirigible, were made by Chaize (France),
by Thunder and Colt (UK) and Peer Lindstrand, in Oswestry (UK). The Radeau
(raft) and the Luge (sled) were constructed by Groupe Solvay (Belgium). The Bulle
des Cimes (bubble of the tree-tops) inflated with helium. The Icos was made by
students from various French technical colleges.

Once the equipment has arrived on site it can be assembled in a fortnight.
However this is only at the end of the 2 to 3 months which it takes to deliver all the
equipment for the expedition in sea freight containers. Time is also needed for the
construction of a ground camp containing dormitories, a restaurant, a kitchen, a
laboratory, an infirmary, washrooms and a landing field for the dirigible.

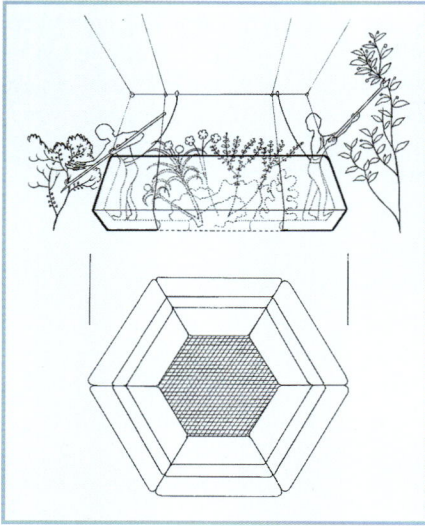

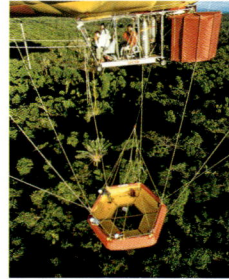

Figure 3 (left) Luge. Figure 4 (below) A harpoon gun fires a line across the forest, enabling a 'canopy bubble' to lift a researcher to explore the tree crown at ease, whilst attached to a rope.

Useful points

The success of the operation depends on the motivation and talent of the team members; it also rests on the logistical aspects of the expedition which can at first appear trivial but which gain increasing importance during a long stay in a tropical forest. I would say for example that it is essential to have:

- a system of high performance VHF radios
- working GPS
- a generator
- comfortable showers
- hot meals at fixed times

Research conducted using this technique

One of the main activities has been the gathering of baseline inventories of plants and animals. Aside from these inventories research conducted since 1986 has principally focused on the following disciplines:

- morphology and architecture of trees and creepers
- ecophysiology, study of epiphytic plants and their environment
- floral biology – smells and aromas
- comparisons of biochemical activity in the understorey and in the canopy of the forest
- the study of intra-tree genetic variability
- study of gaseous exchange between the forest and the atmosphere
- study of herbivory pressure, mapping of insect societies in the canopy, comparison of the diurnal and nocturnal activity of insects and of their activity in the understorey and in the canopy
- observations of arboreal mammals

Using the Radeau des Cimes does not allow prolonged research at the same site as our field operations only last two months. On the other hand, this technique does allow data collection over vast forest areas which is a great advantage. This fundamentally mobile method lends itself well to large-scale comparisons e.g. one site to another in the same forest, one forest to another or one continent to another. It allows a rapid evaluation of the biological diversity and at the same time allows one to take into account the intrinsic variability and characteristics of tropical forests.

The mobility of this technique has been exploited to the full. The successive operations that have taken place are as follows:

1986	French Guiana (Crique Couleuvre)
1989	French Guiana (Petit Saut)
1991	Cameroon (Campo Reserve)
1996	French Guiana (Paracou)
1996	Gabon (Forest of the Abeilles)
2001	Madagascar (Masaola penninsular)

The results of these operations are grouped in expedition reports entitled: Biology of the canopy of an equatorial forest n° I(1990), II(1992), III (1998), IV(2000), et V

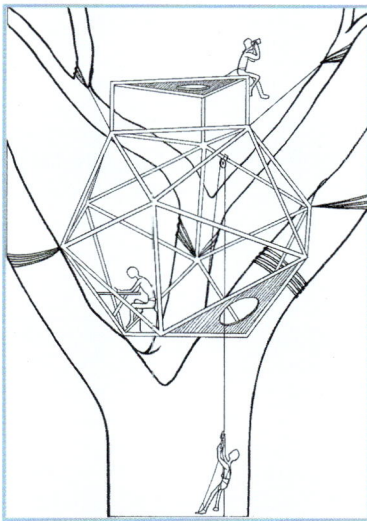

Figure 5 Line Drawing of Icos

(in preparation). These reports are sent on demand to all the scientists who have taken part in the operations. I should also add that in the base camp of the Radeau each evening we organise an informal scientific discussion, which is an excellent means to integrate all the research.

Each of our operations brings together scientists from several nationalities (10 to 14). Margaret Lowman participated in our operations in 1991 in Cameroon and in 1996 in Guiana. I was also co-chairman of the first two international conferences on canopies which took place in the Selby Botanical Gardens, in Sarasota (Florida).

Although there is a great spirit of co-operation between researchers working on tropical canopies, there is no research protocol that is common to all the teams. The fixed methods (walkways, cranes) and the mobile methods (dirigible, radeau) are complementary. However under these conditions, it appears preferable to think collectively about complementary research programs.

Safety

The team of the Radeau des Cimes, in 15 years of operation, has not recorded a single accident or serious incident. This is thanks to a series of preventative measures which are strictly adhered to and which are the responsibility of the expedition leader.

- all researchers wishing access to the canopy are trained by a qualified instructor in traditional techniques of ascending by rope and in the use of modern equipment (e.g. baldric, carabiners, pedals, jumards, descenders)

- whilst in the canopy using any of the techniques (luge, radeau, Icos), tying on is always mandatory, day and night
- wearing a helmet is obligatory on the ground, underneath the equipment (Radeau, Icos)
- a doctor is permanently present in the camp

The range of equipment that we have developed is flexible and adaptable, 'versatile' as the English would say. Thus if a forest is placed at our disposal, we would find or make apparatus that would allow us to explore the canopy. Finally safety is assured by the climbers who are there for use by the scientists. Use of the Radeau, the Luge, the Bulle des Cimes has so far been made in complete safety.

Costs

Operation of this equipment necessitates use of a large team made up as follows: expedition leader; pilot; architect; co-pilot; driver; logistical support person; three assistants for the Radeau, luge and Icos and 5 climbers to help the scientists and to ensure their safety.

The Radeau des Cimes is an economic method, partly because the research only lasts two months and partly because the equipment is not heavy being basically made of cloth. Since 1986, we have spent 3.7 million Euros, included the construction of the equipment, transport, daily logistics and the team salaries during field operations. The construction of the Radeau in 2001 represented an expenditure of 122,000 Euros. Maintenance consists of regular servicing of the dirigible's motor, repair of the pneumatic beams when worn and any other items as required. When not in use the apparatus is stored in Europe. Outside operating periods, maintenance is limited to a flight of the dirigible every three months to keep it in working order.

Funding

Over the last 15 years, our financial resources have come from the following scientific and governmental institutions: UNESCO, CEE, CNRS, various universities, Guianan Space Centre and various regional councils; and private enterprises, principally; Rhône Poulenc, Antargaz, Elf, Shell, Dentsu Prox, Atochem, Dupont, Givaudan, Ricoh and Solvay.

Contact details

Francis Hallé, Professor of Tropical Botany, 109 rue de Lodéve, 34070 Montpellier, France.
Email: halle@radeau-des-cimes.com;
Dany Cleyet Marrel, designer and pilot of the airship
Gilles Ebersolt, architect and inventor of the 'Radeau des Cimes'
www.radeau-des-cimes.com

"Project Hornbill" and Future Lighter Than Air Platforms

By
Graham
Dorrington

PROJECT Hornbill was a venture set up in 1994. It resulted in the development and successful operation of a 400 cubic metre helium filled airship. The first operation was over the forest canopy near the Danum Valley Field Centre (DVFC), Sabah, Borneo, in 1995.

Choice of location

Early in the project it was decided that operations would be restricted to locations where supra-canopy wind speeds were consistently and reliably no worse than a "light breeze" (less than about 4 knots) for periods of at least 2 hours – usually near sunrise or sunset. DVFC was selected because these wind conditions were considered probable, and because it offered relatively good support infrastructure. Helium gas (99%+ industrial purity) was required to fill the airship, and although this would be difficult to obtain in some remote areas, it was relatively easy to obtain and to transport (by truck in cylinders) to DVFC. Authority to operate the airship was granted by the Malaysian DCA – following test flights in the UK (see Figure 1).

Figure 1 "D-4" Airship in flight during "Project Hornbill." Note: the small cylinder above the canoe-like structure beneath the envelope is an insect suction trap. Photo: Graham Dorrington.

Equipment and operations

The "D-4" airship (or dirigible) used for Project Hornbill was developed by Flight Exploration (founded by G. Dorrington). The overall technical aim of the project was to demonstrate that a relatively small, mobile, lighter-than-air platform could be easily used for supra-canopy exploration over wider areas than offered by cranes. The airship's envelope (produced in-house) had a carefully selected streamlined form about 20 m in length and 6 metres in diameter, which is relatively small by airship standards. The lifetime of the envelope was estimated to be about 1 year unsheltered in a tropical rain forest environment. The 400 cubic metre helium capacity permitted a payload of about 75–150 kg (depending on range) to be lifted, i.e. the airship was capable of carrying two people: a pilot and a scientist. An electric-driven propeller system was selected in order to reduce propulsive noise (and hence minimise temporal disturbance of fauna), although this did restrict the operating range of the airship to about 2 km. Controlled hovering flight – within about 5 m of the canopy – was proven to be possible in a validation program comprising 30 successful flights. One deliberate soft-landing was even made onto the canopy (on a flowering Merbau

tree) in still-air conditions and the airship remained motionless for about half-an-hour permitting some close flora observations. The longest flight made was just over 2 hours flying out from DVFC 3 km above the Segama river and returning to a forest clearing measuring 50 m by 50 m. The critical aspect of flight operations was random wind gusting – strongly effecting the ability to hold the position of the airship, and raising the possibility of canopy strikes/entanglement. This problem was avoided by the addition of small electric powered fans, but it was decided after the validation program that further development work was needed.

Research opportunities

Small, mobile LTA platforms are probably best suited to supra-canopy science, e.g., atmospheric sampling, entomological collection and remote sensing missions[1]. They might also be used for upper/outer canopy flora collection/sampling, hence canopy pollination studies appear to be well-suited to the use of this technique. Nest counting, and other zoological surveys using airships with ranges of about 10 km may also be worth considering.

Project Hornbill was a validation program intended to help prove that such LTA platforms are feasible, and although some minor scientific tasks were undertaken, no long-term scientific research program has yet been planned.

Safety

LTA platforms can probably only be used safely when and where still or light breeze conditions can be guaranteed in the nominal operating area for periods of at least 2 hours – preferably during daylight. Use of an anemometer "warning network" around the operating area is recommended. Abseiling equipment should also be carried at all times. In the event of canopy strike/entanglement, it should be noted that sufficient time ought to be available to descend safely, since a properly designed envelope will probably take several minutes to deflate – even in the event of a major envelope rupture. An airship like D-4 can be operated with a team of 3 people, but it would be wise to also have an experienced tree climber on hand. Good ground-air communications are also essential.

Costs and future funding

The project was funded from various sources – including an indirect grant from the Royal Society (London). The total direct project cost was constrained to about $35,000 (1995, US dollars) excluding donations.

A large fraction of the operating costs of Project Hornbill was devoted to the purchase of helium gas. As well as filling the envelope initially, helium lost through venting/leakage had to be replaced. During Project Hornbill about 600 cubic metres of helium gas (90 cylinders in total) was required over 3 months. Helium prices vary widely, but a figure of about $12 per cubic metre provides a rough indication of the helium price at the time. The packaged airship had a mass of about 700 kg (fully packaged for transport to a site), but fortunately it was freighted free-of-charge by Malaysian Airlines.

Looking to the future, no airships like D-4 are commercially available at present, but it's possible that one might be developed for about US$25–50,000, complying with FAA Part 103 Rules, or some similar aviation standard. Flight Exploration is currently developing new (one person and two person) LTA platforms, but these are not intended to be sold, and will only be used for specific scientific projects. Given the considerable success of Project Hornbill with a limited budget (a tiny fraction of the investment in crane systems), future funding has strong justification.

Contact details

Dr. Graham Dorrington
Department of Engineering, Queen Mary, University of London
London E1 4NS UK Tel: 0044 20 7882 5625. Email: g.dorrington@qmw.ac.uk.

1 Note Mitchell advocated the possible use of LTA platforms for canopy research in 1986, see Mitchell, A.W., "The Enchanted Canopy, Secrets of the Rainforest Canopy" pp. 239–245, Collins, London, 1986

Crown Balloon

By
Frank Hodgson

THE crown balloon system is a new method designed to permit routine access to inaccessible regions. This unique system utilizes groups of balloons with two main units having six balloons each, together with two separate but attached balloons, to give a system total of fourteen balloons, each of which is about 9 m in diameter. The six balloons of a main unit are clustered to give either a flat set of six in a triangular grouping or a set of three over three. The inner three balloons are attached to and support a triangular frame (Figure 1).

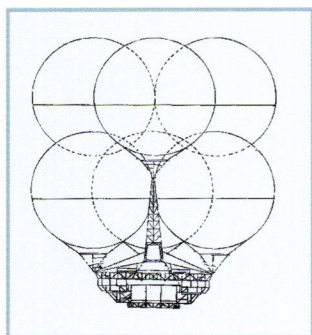

Figure 1 A side view of a main unit stacked 3 over 3

A position halfway between the two extremes of flat or stacked configuration, the "crown configuration", seems to be the most convenient arrangement for travel in general. Movement occurs by throwing out grapples (and/or sea anchors) and then by winching to these points. When travelling downwind, the winches can be used as generators. During daylight hours, power is generated from sheets of photovoltaic cells on the tops of the balloons. The main units also have rechargeable batteries. Free flight is reserved for emergency situations.

The gross lift of the system is about 4,450 kilos and the maximum recommended onboard-crew is sixteen (the minimum is four). All basic life support systems are integrated into the design. A crew should be able to reside over a jungle canopy or over a sub-artic forested region for up to a year. Periodic re-supply by helicopter is suggested to reduce the need to transport large quantities of basic food supplies and to permit transport of a maximum amount of scientific equipment.

The main standard systems on board include; compressed air, electrical power (230VAC, 24 VDC), a vacuum system, refrigeration, cryogenics (liquid nitrogen and liquid helium), air conditioning, heating, a tiny galley, a toilet and shower, a hammer-headed saw (a chain saw for cutting into a jungle canopy to avoid high winds and also to free fouled grapples), a distillation unit for ballast water (from rain water or ground water) and safety nets.

The two separate but tethered balloons are used to set and retrieve grapples and to inspect, clean, and repair each other and the balloons on the main units. They each have a hammer-headed saw, a steam gun, a line winch (for moving to and away from the main units) and communications equipment. Inspection, cleaning, and repair of the balloons is facilitated by having internal lines which maintain the balloon in a spherical shape and by having external lines that permit the rotation of a balloon about one of several axes. The gross lift of a single balloon is about 760 pounds. Each main unit carries 100 pounds of liquid helium, which is freely suspended outboard of one corner of the triangular frame. This amount of helium is enough to reinflate all of the balloons once; the helium boil-off is used as makeup gas. The frames and all components are designed to be easy to disassemble and transport by cargo plane, truck, and/or ship to another site.

This access device can be used by canopy ecologists to facilitate long-term studies within the forest canopy. For instance, the routine collection of foliage from the tops of rainforest canopies will allow an analysis of the mineral content of the leaves, which can then be compared to the mineral content of the soil near the roots of the trees. The allocation of minerals throughout the forest canopy can be easily assessed.

The "crown balloon" system has not been physically constructed. Most of the basic engineering has been completed and the engineering which remains is well-understood. It is difficult to foresee all of the potential applications. It is believed, however, that the system is needed and that once in use it will provide a reliable platform for work in many different environments. The Snow Water Corporation is eager to correspond with canopy researchers who have ideas for research applications or questions regarding this canopy access system.

Contact details

Frank Hodgson, Snow Water Corporation; Phone: (800) 872-5244; AKILO55@Yahoo.com; www.AKILO.com.

Part 2

Methods of
study
in forest canopies

5

Plants

Methods for Sampling Canopy-Dwelling Plants in Forest Ecosystems

By
alini M. Nadkarni

THE upper tree canopy of many forest ecosystems fosters extremely diverse plant communities, which include vascular and non-vascular epiphytes, hemi-epiphytes, and parasites. Canopy-dwelling plants contribute substantially to overall forest biodiversity and biocomplexity by providing resources for arboreal vertebrates, invertebrates, and microbes, and by participating in nutrient and water cycling, and gas and energy exchanges. Trees as the most obvious structural component of forest canopies, have trunks, branches, and leaves that constitute the canopy infrastructure and provide mechanical support for thousands of species of arboreal plants and animals.

Much has been published on canopy plants (e.g. Madison 1977, Kress 1986, Benzing 1990, Putz & Mooney 1991, Rhoades 1995), but protocols for sampling and analyzing their distribution and abundance are still scattered and inconsistent. In this review, I describe some of the most common methods for sampling canopy-dwelling plants, with an emphasis on the epiphytic vascular plants, epiphytic cryptogams (non-vascular plants that include lichens and bryophytes), primary and secondary hemi-epiphytic vascular plants (lianas and vines), and arboreal parasitic vascular plants. Arboreal fungi and free-living algae are so poorly known that there is little to review.

Rather than an exhaustive review of all papers that involve the sampling of canopy plants, I have taken selected papers that represent five major categories of canopy ecological studies (general distribution and composition; canopy plant/substrate interactions; phenology; nutrient cycling; and forest management). For each category, I have selected 3–10 papers that encompass the variability of different sampling methods. I have excluded studies dealing with the taxonomy, physiology, and genetics of canopy-dwelling plants, and focused on those that deal with field ecological questions. The original papers are cited here and should be used by the reader for details on methodology.

Review of methods
General distribution and composition

The most basic way of understanding canopy plant distribution is by enumerating the flora on a single tree or a few trees of the same species. For example, the epiphytic flora of a single old Huon pine in Tasmania was sampled to document the number of epiphytes and their zonation (Jarman & Kantvilas 1995). The tree

was divided into five height classes, and any parts of the tree that had fallen were 'reconstructed' and included in the tally. Similarly, Cornelissen and Ter Steege (1989) differentiated six height zones, and five trees of the same species were then climbed. In each plot within the trees, the total cover percentages of different functional plant groups were estimated and substratum characteristics recorded.

At a larger spatial scale, Wright *et al* (1997) enumerated the trees and lianas on a one-hectare plot in Papua New Guinea by establishing transects of varying sizes to assess tree and liana size and location. Because not all trees were climbed, the researchers noted that liana species richness may have been underestimated. In Brazil, Fontoura (1995) documented the distribution of five species of bromeliads on phorophytes. In 20 sample plots, she observed the height, phorophyte species, and frequency of bromeliads, and carried out univariate and multivariate analyses to detect relationships between host and epiphyte.

Many studies concerned with the distribution of canopy plants over large areas or altitudinal gradients establish long transects and then sample epiphytes near ground level. For example, a study in Tuscany, Italy of epiphytic lichens examined them on the lower trunks of a single host tree species over an elevational gradient from 0 to 900 m, using small quadrats (30 x 50 cm) as a sampling unit (Loppi *et al* 1997). In contrast, other studies have used canopy access techniques to sample tree crowns *in situ*. Hietz & Hietz-Siefert (1995), for example, ascended trees and took samples from six zones within the canopy. They obtained both branch dimensions and quantitative estimates of cover and were thus able to document relationships between microhabitat characteristics and different species of epiphytes.

Canopy plant/substrate interactions (including host-tree specificity)

Biogeographical and ecological studies on canopy plants indicate that there are two major factors that control their distribution: environmental conditions (e.g. moisture availability, temperature, light intensity); and substrate features (e.g. bark pH, rugosity, and water-holding capacity) (Wolf 1994). A major issue in canopy plant/substrate studies is whether canopy plants exhibit host tree specificity. Most studies investigate whether certain canopy plants 'prefer' certain species of host trees, but the question of whether certain host trees foster particular communities or species of epiphytes has also raised interest (e.g. Barclay-Estrup & Sims 1978).

Some studies have focused on the factors that control the distribution of canopy plant associations over altitudinal, successional, or other environmental gradients. One study (Wolf 1994) examined the epiphyte communities along a transect (1000–4130 m) in the Andes. He selected four host trees at each of 15 sites, and used the Braun-Blanquet approach in phytosociology to establish relevés from ground to outer canopy. Species cover (in percentages) was recorded and analyzed with a phytosociological analysis of all taxa and a canonical correspondence analysis. Host tree specificity was tested in a Monte Carlo permutation procedure.

In a study that took advantage of massive treefalls following a windstorm, Eversman *et al* (1987) sampled 15 fallen trees of three different species, recording the presence and location of epiphytic species from the base to the tip of each tree. This yielded data on the microdistribution of epiphytic lichens throughout the tree. Bark was also sampled, and taken to the lab to determine water-holding capacity and pH. These characteristics appeared to explain the interspecific difference of lichen distribution on the different host trees.

More focused studies have been done on individual species. For example, Ackerman *et al* (1989), established plots (3 x 30 m) in areas where the orchid *Encyclia krugii* is common. All trees and shrubs were identified, and the presence of the orchid was noted. Phorophyte physical characteristics were judged as either rough or smooth.

Another study in a southern deciduous hardwood forest (Talley *et al* 1996) showed that a root-climbing liana, *Rhus radicans*, is not distributed randomly among potential host trees species. Based on surveys of the distribution and abundance of host trees in 30 contiguous 20 x 20 m quadrats, the researchers tallied the number of vines on each tree. They noted that there are different ways to calculate the 'relative abundance' of hosts. If they use the number of stems available, as the proportional contribution of the number of stems in the forest, this ignores the fact that bigger trees are bigger targets for colonization. Instead, they used basal circumference as a measure of the abundance of potential hosts. They then carried out a c2 test to test for significant divergence between actual vs. potential colonization of the liana.

Nalini Nadkarni in search of canopy epiphytes using single rope technique. Photo courtesy of Nalini Nadkarni

Phenology

Phenology refers to the seasonality of plant life history traits, and has been measured most frequently for the timing of leaf emergence, flowering, and fruiting, especially for pollination and fruit dispersal studies. A subset of canopy plants is selected for their accessibility and visibility either from the ground or in the trees, and at intervals ranging from one week to several months, the researcher records the timing of the plant activity of interest. Efforts are made to disperse the distribution of sample plants among different trees (e.g. Putz *et al* 1995), but some studies have been restricted to just one or several trees (e.g. Ackerman *et al* 1997.).

Nutrient cycling

Canopy-dwelling plants have been the object of nutrient cycling studies in a variety of temperate, tropical, and boreal ecosystems (Coxson & Nadkarni 1995). Because epiphytes derive their nutrients from atmospheric sources (Nadkarni & Matelson 1992), they make excellent subjects to study the sources and sinks of nutrients in rain, mist, and dry deposition.

In general, canopy nutrient cycling studies have involved the assessment of biomass and nutrient content held within the canopy, using epiphytes and their associated dead organic matter collected from a variety of trees and canopy microsites. These range from one or a handful of trees using single-rope technique (e.g. Nadkarni 1985) to sampling from multiple crowns, and the use of felled trees (e.g. Eversman *et al* 1987). Subsamples of epiphytes are then separated into components (e.g. live vascular plants, cryptogams, dead organic matter), and then

extrapolated to a branch- and tree- level. The stand-level distribution of trees is measured (usually placed into size classes), which is used to extrapolate the epiphyte biomass to a whole forest level.

An alternative method of estimating epiphyte biomass was put forward by McCune (1994), using fallen epiphyte litter on the forest floor. This was effective and accurate in the tall temperate coniferous forests where he worked, and presented a viable alternative to tree-climbing for the assessment of lichens. However, bryophyte litter tends to be highly aggregated, and is subject to greater error because of the variable distribution patterns.

Other studies have considered the effects of epiphytes on nutrient transfer via throughfall and leaching. These effects have been assessed in the field with 'branch frames' (funnels held beneath branches), and by taking the epiphytes from the field and conducting leaching experiments in the laboratory (Clark *et al* 1998).

Forest management

Epiphytes, particularly non-vascular epiphytes such as lichens and bryophytes, are known to be decreasing because of habitat destruction, forest operations, and air pollution (Bernes 1994). Many of these species are considered unable to tolerate the drastic changes in light conditions and air humidity that accompany clear-cutting. Their poor dispersal abilities also make them vulnerable to forest management activities.

Some studies have taken a descriptive approach (e.g. Esseen & Renhorn 1996) to understanding the patterns that regulate the occurrence and dynamics of species in undisturbed vs. managed (selectively logged) stands. They selected stands and established plots (2 ha in area) that avoided edges. Transects (200 m long) were then set up, and at 20 m intervals, trees were measured for size. A single branch 1–4 m in height was selected per tree , and the branch was removed, placed in a large plastic bag, and transported to the lab to determine lichen composition and biomass.

Other studies have taken an experimental approach, using epiphyte transplants on trees (or patches of trees) that have been isolated due to tree-felling and other forest operations (McCune *et al* 1996). For example, Hazell & Gustafsson (1999) selected 100 clear-felled sites adjacent to mature forests, and took a random subsample of these areas. The aspen trees they used were at least 20 m from the forest edge; they used 280 'receiver' trees, 76 of which were clustered, 64 scattered, and 140 in forest (as controls). Transplants of two species (one lichen, one moss) were collected from stems at two times of the year, and uniform squared pieces (6 cm^2) were transplanted on the north and south sides of stems of receiver trees 150 cm from the ground. A plastic net was applied with metal staples, and sprayed with water. Survival was evaluated by categorizing the transplants into four categories (total survival to total death); vitality was assessed in five categories by noting the remaining living material of the shoots/thalli.

Summary and conclusions

One of the major obstacles in quantifying the biomass, distribution, and composition of canopy plants and their ecological interactions with their substrate and other organisms has been the lack of consistency in sampling protocols. Even as limited a review of techniques as presented here reveals an astonishing diversity of protocols and methods; few studies or sites can be directly and effectively compared.

Some of the issues of sampling arise from the fact that tree surfaces and substrates are extremely topologically and microclimatically complex. Although human beings tend to see a single tree as an individual (hence N=1), the multiple angles, size classes, rugosity, and exposure to microclimatic variability of branches within a single tree constitute an enormous range. Very few studies have assessed the within- tree versus between-tree levels of variability in a statistically rigorous manner. This is needed if canopy researchers are to establish protocols that can be used reliably across habitats.

The obvious problems of accessibility also make statistically sound sampling difficult. Nearly all researchers admit to some degree of bias in their choice of trees; they must either be of the appropriate size and architecture to climb safely, or occur directly below a canopy crane or canopy sled. Although the latter two techniques have allowed researchers to sample a much larger number of trees, and a greater variety of size classes, studies are still restricted to a finite and potentially biased set of trees without replication of stands.

The increasing number of studies on canopy plants promises to make the issues of development of harmonized methods increasingly important. Communication of the techniques – both before and after studies are completed – will be essential to make this occur efficiently and effectively.

Contact details

Dr Nalini M. Nadkarni, The Evergreen State College, Olympia, Washington 98505 USA
Tel: (360) 867–6621, Fax: (360) 866–6794, Email: nadkarnn@evergreen.edu.

Literature cited

Ackerman J, Montalvo A & Vera A (1989). Epiphyte host specificity of *Encyclia krugii*, a Puerto Rican endemic orchid. *Lindleyana* 4:74–77.

Ackerman J, Meléndez-Ackerman E & Salguero-Faria J (1997). Variation in pollinator abundance and selection on fragrance phenotypes in an epiphytic orchid. *American Journal of Botany* 84:1383–1390.

Barclay-Estrup P & Sims R (1978). Epiphytes on White Elm, *Ulmus amaricana*, near Thunder Bay, Ontario. *Canadian Field-Naturalist* 93:139–143.

Benzing DH (1990). *Vascular epiphytes.* Cambridge University Press, Cambridge.

Berg A, Ehnström B, Gustaffson L, Hallingback T, Jonsell M & Weslien J (1994). Threatened plant, animal, and fungus species in Swedish forests: distribution and habitat associations. *Conservatuon Biology* 8:718–731.

Clark K, Nadkarni N & Gholz H (1998). Growth, net production, litter decomposition, and net nitrogen accumulation by epiphytic bryophytes in a tropical montane forest. *Biotropica* 30:12–23.

Cornelissen J & Ter Steege H (1989). Distribution and ecology of epiphytic bryophytes and lichens in dry evergreen forest of Guyana. *Journal of Tropical Ecology* 5:131–150.

Coxson D & Nadkarni N (1995). *Epiphytes and nutrient cycling.* Pp. 495–543 *In*: N. Lowman and Nadkarni N, eds. Forest Canopies. Academic Press, San Diego, California USA.

Eversman S, Johnson C & Gustaffson D (1987). Vertical distribution of epiphytic lichens on three tree species in Yellowstone National Park. *The Bryologist* 90:212–216.

Esseen P & Renhorn K (1996). Epiphytic lichen biomass in managed and old-growth boreal forests: effect of branch quality. *Ecological Applications* 6:228–238.

Fontoura T (1995). Distribution patterns of five Bromeliaceae genera in Atlantic rainforest, Rio de Janeiro State, Brazil. *Selbyana* 16:79–93.

Hazell P & Gustafsson L (1999). Retention of trees at final harvest – evaluation of a conservation technique using epiphytic bryophyte and lichen transplants. *Biological Conservation* 90:133–142.

Hietz P & Hietz-Seifert U (1995). Structure and ecology of epiphyte communities of a cloud forest in central Veracruz, Mexico. *Journal of Vegetation Science* 6:719–728.

Jarmon S & Kantvilas G (1995P). Epiphytes on an old Huon pine tree (Lagarostrobos franklinii) in a Tasmanian rainforest. *New Zealand Journal of Botany* 33:65–78.

Kress WJ (1986). The systematic distribution of vascular epiphytes: an update. *Selbyana* 9:2–22.

Loppi S, Pririntosos S & De Domincis V (1997). Analysis of the distribution of epphytic lichens on *Quercus pubescens* along an altitudinal gradient in a Mediterranean area (Tuscany, Central Italy). *Israel Journal of Plant Sciences* 45:53–58.

Lowman MD & Nadkarni NM (1995). *Forest Canopies*. Academic Press. San Diego, California.

Madison M (1977). Vascular epiphytes: their systematic occurrence and salient features. *Selbyana* 2:1–13.

McCune B (1994). Using epiphyte litter to estimate epiphyte biomass. *The Bryologist* 97:396–401.

McCune B, Derr C, Muir P, Shrirar A, Sillett S & Daly W (1996). Lichen pendants for transplant and growth experiments. *Lichenologist* 28:161–169.

Nadkarni N (1985). Nutrient capital of canopy epiphytes in an Acer macrophyllum community, Olympic Peninsula, Washington State. *Canadian Journal of Botany* 77:136–142.

Nadkarni N & Matelson T (1992). Biomass and nutrient dynamics of epiphyte litterfall in a neotropical cloud forest, Costa Rica. *Biotropica* 24:24–30.

Putz FE & Mooney HA (1991). *The biology of vines.* Cambridge University Press, Cambridge, UK

Putz R, Romano G & Holbrook N (1995). Comparative phenology of epiphytic and tree-phase strangler figs in a Venezuelan palm savanna. *Biotropica* 27:183–189..

Rhoades FM (1995). Nonvascular epiphytes in forest canopies: Worldwide distribution, abundance, and ecological roles. Pages 353–408 in *Forest canopies*, (Lowman MD & Nadkarni NM eds.). Academic Press, San Diego, California.

Talley S, Lawton R & Setzer W (1996). Host preferences of *Rhus radicans* (Anacardiaceae) in a southern deciduous hardwood forest. *Ecology* 77:1271–1276.

Wolf J (1994). Factors controlling the distribution of vascular and non-vascular epiphytes in the northern Andes. *Vegetatio* 112:15–28.

Wright D, Jessen J, Burke P & de Sliva Garza H (1997). Tree and liana enumeration and diversity on a one-hectare plot in Papua New Guinea. *Biotropica* 29:250–160.

Techniques for the Study and Removal of Canopy Epiphytes

By
Martin Ellwood
and
William Foster

EPIPHYTIC plants are an important component of the niche variety that is fundamental to the maintenance of species diversity in forest canopies. Although epiphytes can be observed and sampled *in situ*, it is often necessary to remove them from the canopy for closer observation, and to experiment with the plants themselves or the organisms associated with them. The techniques we consider here were developed for entomological work on Bird's Nest Ferns (*Asplenium nidus L.*) in the high canopy (above 40 m) of lowland dipterocarp rainforest in Borneo. Bird's Nest Ferns can weigh up to 200 kg and are probably one of the largest epiphytes in the world. However, the following techniques can be modified to accommodate plants of all sizes, growing at any height, in virtually all forest types. In the next section we will describe (a) how precision fogging can be used to sample arthropod communities in epiphytes, (b) an effective protocol for removing epiphytes from the canopy and (c) a method for extracting invertebrate animals from epiphytes.

The techniques
Precision fogging

Precision fogging can be used to sample any of the discrete components of rainforest canopies, and providing that lines can be inserted and positioned accurately from the ground, there is no need to climb trees. Unlike conventional fogging, trays are raised to within only a few metres of the target habitat and the fog is concentrated in a particular area. Individual regions, and different layers of the canopy, can be targeted precisely, and these samples provide a useful context for more focussed studies. The accuracy of precision fogging is far greater than the conventional technique because the fog is more focussed and animals drop only a few metres, thus reducing a drift of potentially many metres (Floren & Linsenmair 2000). Any number of trays can be raised to the canopy providing that enough lines can be inserted. The number of trays that can be attached per line will depend upon tray design and weight limits.

To sample an epiphyte by precision fogging, the aerial tray line should be positioned over the supporting branch and central to the plant. Trays of the same area as the plant should be raised to *ca* 1 m of its base, to minimise drift and make a precise collection (Figure 1). If the tray area exceeds that of the plant, the sample will be contaminated with animals from the canopy above. Similarly, if the epiphyte is to be removed and sampled in greater detail after fogging, additional trays should be placed *ca* 1m above it during fogging so that crown animals are not intercepted by the plant itself. These trays provide a useful sample of the canopy directly above the epiphyte.

The fogger should be raised into the canopy in the normal way using a pulley system. Care should be taken to avoid tangling the fogger support ropes, or burning them with the hot exhaust nozzle. These dangers can be alleviated by placing the fogger line over a branch at least five metres above the epiphyte so that the

Figure 1 Capture tray positioning underneath a specific epiphyte and on the forest floor

support ropes are kept apart (Figure 2). Trays should be left for two hours after fogging to ensure that as many animals are collected as possible. The location of each tray should be recorded and the animals sorted separately.

Removing epiphytes

For removing epiphytes, two lines must be inserted into the canopy (see section 1). The first is for placing the climbing rope, the second for lowering the epiphyte. Tree architecture and height requirements will determine which line-insertion technique is the most suitable. Each of the lines should be inserted at least five metres above the epiphyte. Failure to insert the lines at the correct height restricts

Figure 2 Fogger operating in the canopy. Photo: Martin Ellwood.

lateral movement at the ends of the rope. Placing lines over branches well above the working area also reduces the risk of entanglement of people and equipment. The rope used for lowering the epiphyte must be positioned correctly, because it should be under tension before the fern is removed. If there is slack in the rope, the epiphyte can free-fall and be dropped or damaged. The application of tension to the rope before the plant is removed can only be accomplished if the rope is positioned higher than the epiphyte. As the plant is lowered, lateral movement is also essential if it is to avoid becoming entangled in the understorey.

Epiphytes should be enclosed in mosquito netting before they are removed to prevent the loss of animals or leaf-litter as they are lowered and carried from the forest. The netting should be large enough to enclose the epiphyte completely without needing to be stretched. The netting process must be carried out quickly and efficiently otherwise animals will be lost, although if the plant is being removed after an insecticide fog many animals will remain under sedation. Small epiphytes can be netted quite easily, but for larger plants the net must be cast from above, in the same way as a fishing net. Once the net has been draped over the entire plant, a loop of rope should be passed around its base and fastened using a slip-knot. Hence, as the rope is pulled the loop tightens and the net is sealed. The plant should then be attached to the lowering rope, again using loops and slip-knots. As the plant is then lifted from the branch, both the netting and the support ropes tighten automatically. A guide rope should be attached to one of the support ropes so that the epiphyte can be manoeuvred from side to side as it is lowered. This helps to prevent the plant becoming entangled in the understorey as it descends. Once on the ground, the epiphyte should be secured in a protective waterproof sheet so that it can be carried safely.

Extracting animals from epiphytes

The protective sheet in which the epiphyte is carried should be white so that when the ferns are uncovered and the sheet unfolded, animals running across the surface will be more conspicuous. After being taken from the forest, the epiphytes should be fogged again as a precaution to prevent animals from escaping when the protective sheet is removed. It is likely that the fog will not affect all animals and therefore the plant should be unwrapped slowly and in the presence of several enthusiastic assistants armed with forceps and collecting jars. The white sheet, which should be unfolded and laid flat on the ground, creates an ideal surface for collecting, and brushes are a useful way of sweeping up animals.

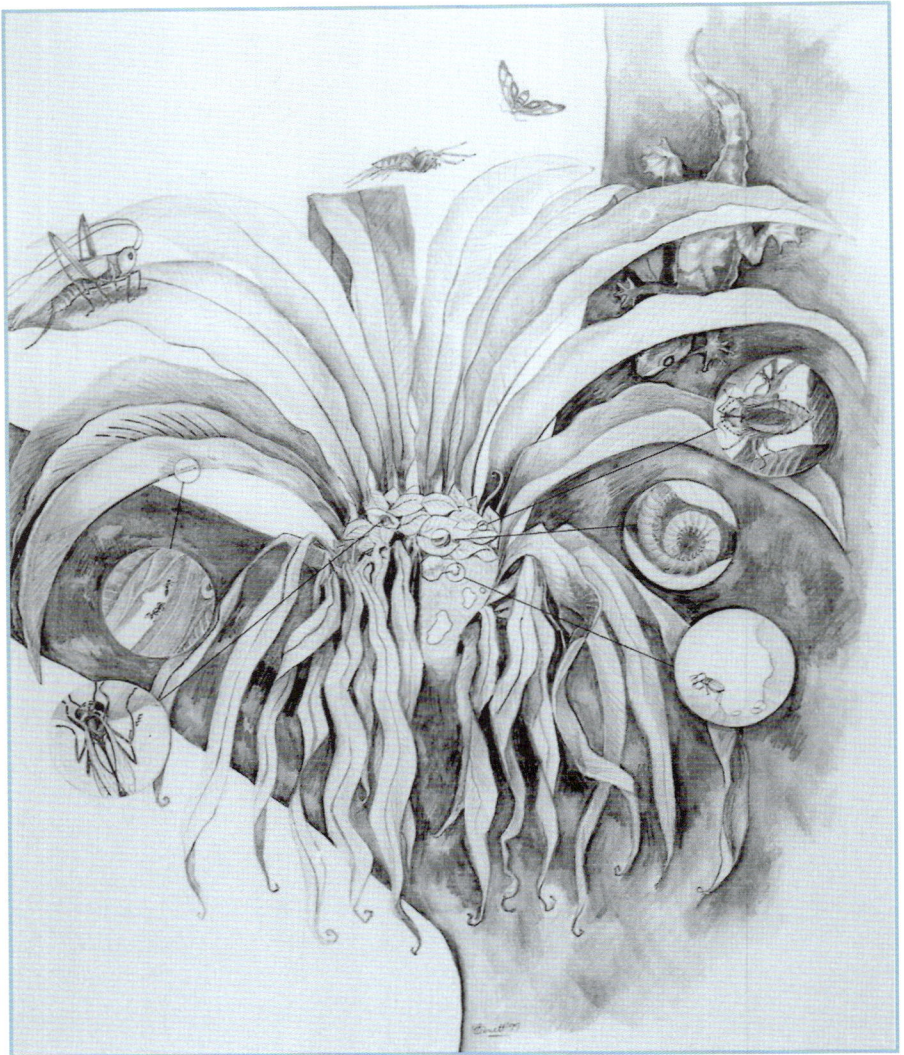

Figure 3
Large leaves cut
away from Bird's
Nest Fern to
enable access to
components

In some cases it may be necessary to determine how many animals were in the epiphyte as larvae. To do this, a small part of the plant should be netted and placed under the original growing conditions to allow the larvae to develop and emerge as adults. Once these adults have emerged, the plant can be sampled in the same way as the rest of the epiphyte.

When sampling epiphytes, a useful approach is to divide them up anatomically. The main structures such as leaves, stems and roots are then kept separate, as this makes it possible to determine which part of the plant the animals came from. With Bird's Nest Ferns, the order of operation begins with the removal of the leaf-litter, then the large waxy leaves are cut away, followed by the dead leaves (Figure 3). Finally the remaining mass of organic soil and roots is dissected. Pruning saws and garden clippers are the most useful dissection tools, as many epiphytes contain lignin in their roots and stems making them very tough to cut. As components of the plant are removed, they should be weighed if necessary and placed directly into Winkler bags.

Winkler bags are an effective way to remove invertebrate animals from plant material, organic soil and leaf-litter. In some cases, the size of these bags may need to be increased to accommodate larger plants. We use large (2m x 1m) bags made of white cotton to collect animals from Bird's Nest ferns. A cylindrical wire basket, which is smaller than the bags and does not touch their sides, is placed inside each bag. This design allows the circulation of air, which accelerates the desiccation process. A container with 70% alcohol inside should be suspended under each bag to collect the animals as they emerge. These containers must be monitored closely and changed when they are between one third and one half full of animals, otherwise the alcohol will become dilute and the animals may rot, especially in high temperatures. With Bird's Nest ferns, these containers may need to be changed as often as two or three times per day. When animals stop emerging from the bags, the plant material should be sorted by hand under a magnifying lens, and examined under a microscope, to ensure that no animals remain. At this point, plant dry weights can be measured.

Using natural desiccation to extract animals does not require electricity, which means that this technique can be used at remote field sites. Some samples taken from Winkler bags will be clean, whereas others will contain silt or debris. Animals most at risk from damage should be processed first, by separation from plant material and debris and by being placed in labelled vials. Some preliminary sorting to taxa can also be accomplished at this stage, depending on time constraints. Larvae should be separated and stored in the same way as adults. In large epiphytes, the collection may grow into many hundreds of vials. The alcohol needs to be inspected and changed regularly. Once in clean alcohol, the samples can be stored and transported safely. Ultimately the collection, in clean alcohol at a stable 70%, should be placed in frozen storage.

Contact details

Martin D.F. Ellwood, The Insect Room, Department of Zoology, University of Cambridge, Downing Street, Cambridge CB2 3EJ, United Kingdom. Tel: 01223 331768 (Direct Line). International: +44 1223 331768. Fax: 01223 336676. International: +44 1223 336676. email: mdfe2@cam.ac.uk

Literature cited

Basset Y (2001). Invertebrates in the canopy of tropical rainforest – how much do we really know? *Plant Ecology* 153: 87–107.

Besuchet C, Burckhardt D & Lobl I (1987). The Winkler / Moczarski eclector as an efficient extractor for fungus and litter Coleoptera. *The Coleopterists' Bulletin* 41: 392–394.

Compton SG, Ellwood MDF, Davis AJ & Welch K (2000). The flight heights of chalcid wasps (Hymenoptera: Chalcidoidea) in a lowland Bornean rain forest: fig wasps are the high fliers. *Biotropica* 32: 515–522.

Ellwood MDF & Foster WA (2001). Line insertion techniques for the study of high forest canopies. *Selbyana* 22: 97–102.

Floren A & Linsenmair KE (2000). Do ant mosaics exist in pristine lowland rain forests? *Oecologia* 123: 129–137.

Stork NE & Brendell MJD (1993). Arthropod abundance in lowland rain forest of Seram. *In Natural History of Seram, Maluku, Indonesia* (Eds, Edwards ID, Macdonald AA & Proctor J) Intercept Press, Andover, pp. 115–130.

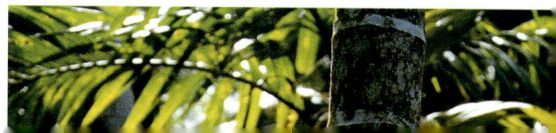

Techniques for Spatial Data Analysis in Epiphyte Ecology

By
Hein JF VanDunné
and
Jan HD Wolf

THE development of new canopy access tools has led to an increase in canopy research (Gradstein *et al* 1996, Lowman & Wittman 1996). The description of floristic composition of epiphytic vegetation has been facilitated by better access, especially in areas where it is not easy to find expert climbers. However, for studies relating epiphyte composition to structural aspects of the forest, canopy access alone is not enough. Without appropriate tools for measurement and data analysis, studying the development of epiphytic vegetation through time is difficult. Techniques are needed which allow the researcher to partition out the effects of availability of substrate on the overall geometric pattern of the epiphytes under study.

The space occupied by epiphytes in forest canopies has been divided into qualitative zones (Johansson 1974) or 'spheres' (Freiberg 1996). Johansson's system of zones has been frequently applied, and is useful for a general interpretation of species distributions in trees. For the most part these zones are related to the climatic gradient present on trees, where at the two extremes the outer canopy offers a dry, unstable habitat and the base of the trunk a moist and stable one. This system works well for small-scale studies on one or a few trees of similar height. However on a larger scale, an alternative must be sought because the position of epiphytes is sometimes better described in relation to the forest. For example, the outer canopies of an understory or emergent tree are not readily comparable.

The interpretation of the variation in ecological data sets which can be attributed to space remains contentious, and has led to different schools of thought on ecology (Crawley 1997). Nevertheless, plant communities often have non-random spatial distributions, which can be dynamic, even at small time scales (Herben *et al* 2000). A number of textbooks on numerical ecology highlight the importance of space and offer methods for spatial data analysis (Legendre & Legendre 1998, Young & Young 1998, Dale 2000). Of these textbooks, only Dale (2000) recognizes epiphytes as a special case in which three dimensions should be considered and he then expands briefly on the possible direction of research in this area.

Epiphytes can not occupy every coordinate in the three dimensional volume of the forest, but are constrained by the presence of tree surface. The nature of this constraint can be described and quantified as a function of the availability and suitability of substrate. Availability depends on the spatial distribution of host tree surface in the forest stand. Suitability depends on the position of a particular point in space in the context of the forest stand (environmental conditions) and the characteristics of the bark surface (host preference). In epiphyte studies a very high proportion of the volume of a forest stand (from the forest floor to the upper level of the forest canopy) will be undefined, or non-existent as far as epiphytes are concerned. Of the proportion that is defined, possibly the majority may be unsuitable for epiphytes.

In recent years, many papers on methodology in epiphyte research have appeared (e.g Gradstein *et al* 1996, Shaw & Bergstrom 1997, Bergstrom & Tweedie

1998, Nieder & Zotz 1998). However, most of these emphasize field methods, while perhaps it is the experimental design and statistical analysis of the data that are most difficult. It is also at this latter level that we must try to find some consensus among scientists in the field, so that data in publications become more comparable. Including forest structure in a standardized way would facilitate comparison between forests and between published data sets. One attempt to do so was the publication by Bergstrom & Tweedy (1998). These authors break down the epiphytic habitat into apparently intuitive hierarchies (a forest into trees, branches, twigs etc). In our opinion, however, they over simplify the problem by suggesting that each level in this hierarchy can be used as an independent sampling unit. Two branches of the same tree are not independent and any conclusions can not be extrapolated beyond that individual tree. Using two objects as replicates of each other when they are not, most commonly because they are not independent samples is called pseudoreplication (Hurlbert 1990). This is often difficult to avoid because knowing whether two samples are independent requires some prior knowledge about the spatial relationships in the data set. This information is seldom at hand for epiphytes. However, by placing the study in a spatial context, at least the problems with pseudoreplication which originate in lack of spatial independence can be made visible. Determining the sampling efficiency of a sampling design *a priori* is often difficult, unless the terrain and the taxonomic group under study are well known. Nevertheless, in many cases it would be valuable to have a measure of how much the spatial relationship between the objects contributes to the total variation in the data set (Legendre 1993, Koenig 1999).

A spatially explicit context requires the *a priori* definition of three aspects of the experiment or sampling design. First, the measurement unit has to be defined. These units can be either based on a lattice, so that average values of variables can be calculated in each grid cell, or defined as points (Isaaks & Srivastava 1989, Cressie 1993) (Figure 1). In the latter case, each point has unique attributes (for example, species size and coordinates etc.). Second, the number of dimensions to be included in the study have to be defined, ranging from one (transect study) to four (three dimensional sampling with repeated measurements in time). Finally, these dimensions have to be evaluated to decide whether they are continuous or discontinuous (constrained). An example of continuous spatial dimensions are the (x, y) coordinates of the tree stems because these could have been placed at any point within the sampled plot. An example of a constrained pattern is the position of an epiphyte on a tree. This epiphyte could not have occurred anywhere but on one of the trees inside the plot under study.

Each of the above decisions has implications for the choice of field methods and the possibilities for statistical analysis of the data. The short review of methods below will highlight some of the implications but is far from exhaustive.

Measurement units

In epiphyte ecology, measurement units are usually defined as either an area of fixed size (quadrats or plots) or individual trees (or portions thereof, branch segments being most common). In Figure 1 the spatial scales used in epiphyte studies are divided into spatial units, habitat units, and point patterns, where the coordinate of each unique object is recorded. Both spatial units and point patterns can be extended (at least conceptually) to three dimensions, where the grids cells in

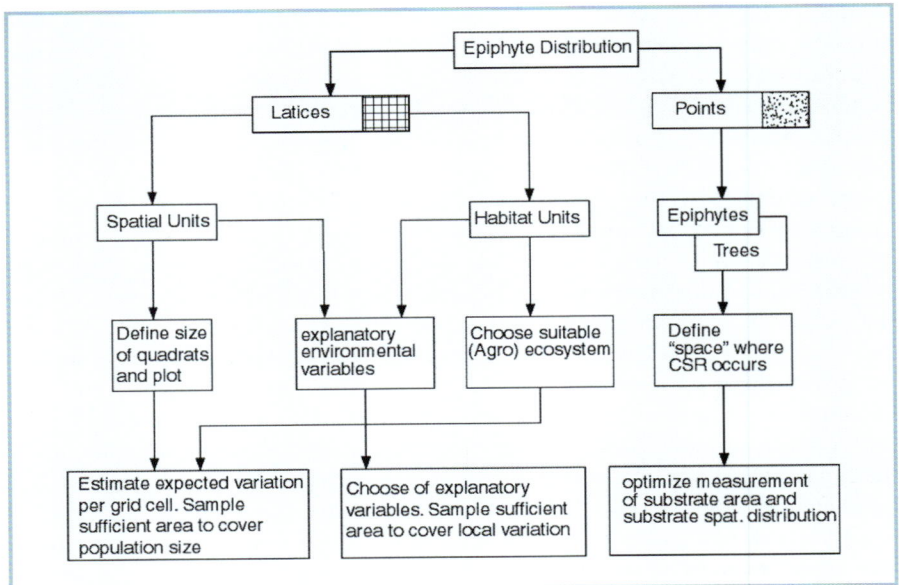

Figure 1 Selection and definition of measurement units

lattice based models are extended to cubes with a given volume, and points are recorded using a coordinate system with three dimensions. However, while the latter option is still feasible in the field, working with cubes of a given volume is more difficult, but some attempts in this direction have been made (van der Meer 1997).

If some form of spatial analysis is to be performed on the data all measurement units need to be geo-referenced. The size of the quadrats and/or the plots have to be defined based on the expected scale of the patterns that the researcher is looking for. Often, analysis is easier when a regular lattice can be used for sampling, but this is not always necessary. In the case of point patterns the main problem for its application to epiphyte ecology is that it is difficult to model complete spatial randomness given the constraints imposed on epiphyte presence by the availability of substrate. To do so, a precise map of all available substrate is required, and standard methods for pattern analysis can not be used.

Plantations offer a system for studying the distribution of epiphytes which allows the definition of habitat units as grid cells for subsequent statistical analysis, instead of spatial grids in a strict sense which are superimposed on a vegetation. Madison (1979) has used this to study the distribution patterns of epiphyte species. His use of the Poisson distribution to test for spatial randomness should be avoided however, as it has been shown to lead to misinterpretations of the data (Hurlbert 1990).

Data analysis
Methods based on ordination

Although some ecologists are critical of multivariate techniques (e.g. Crawley 1997), ordination methods are a useful tool to interpret ecological data. Borcard *et al* (1992) describe a method they call Partial Canonical Analysis, in which the component parts of the variation in the species matrix can be assigned to the spatial and the environmental data as fractions of the total variance and expressed as percentages.

This is an efficient method for working out how much of the variation in the species data is attributable to either space or the other explanatory variables. Superimposing these fractions of variance on a topological map can help in the interpretation of the results (Legendre & Legendre 1998). The matrix is built with components of a polynomial trend-surface analysis which is a special case of the usual multiple regression techniques in which observed values of the variable of interest are regressed against the independent position variables. The resulting line, surface or volume, represents the long-range changing aspects of the spatial variation of the variable of interest. The short-range, random component of the model is represented by the deviations from regression. For further reading on trend-surface analysis, see e.g. Burrough (1987). For trend-surface analysis the same assumptions as with multiple regression apply. If applied strictly, these assumptions would invalidate this method for discontinuous spatial phenomena. It is probably best applied to epiphyte studies were sampling is carried out in spatial units. Also, the method requires sufficient variation between grid cells, and sufficient species variety, to work well.

Geostatistics

Geostatistics is a multidisciplinary field which incorporates elements from geology, statistics and mathematics to model spatial variability (Cressie 1993). Initially aimed at ore-reserve estimation in the mining industry, geostatistical tools for pattern analysis can be successfully applied in ecological studies (Rossi *et al* 1992). Some of these methods can be extended to three dimensions. However, these methods, while useful for spatial prediction, are limited for conditional prediction. If, for example, the abundance of a species is sampled in non-contiguous quadrats, and the presence of that species in the intermediate quadrats needs to be predicted, methods for spatial prediction such as kriging could be used (for further explanation see http://www.geo.ed.ac.uk). In cases where environmental variables need to be taken into account, and the prediction needs to be generalized beyond the sampled area, then a more general form of a regression model is required. For these kind of studies permutation regression (Legendre *et al* 1994) offers more flexibility.

Geo-referenced data is often stored in geographical information systems (GIS). In essence a GIS is a specialised data base developed for flexible storage and retrieval of data with a spatial component. Examples of the use of GIS can be found in Bader *et al* (2000), and VanDunné (in press).

Permutation regression

Permutation regression as described by Legendre *et al* (1994) includes variables in the form of distance matrices, which can either be binary state (one variable in a matrix of distances were the values of that variable in each sampled quadrat is compared to all other sampled quadrats), or multi-state matrices (a number of variables are collapsed using a distance or similarity index) (Legendre & Legendre 1998). This approach allows the inclusion of many environmental variables and is flexible in the way species data are included (VanDunné 2001). The resulting

model tests the effect of geographic distance relative to environmental variables, and allows the researcher to test a spatially explicit ecological hypothesis.

The calculation of a permutational regression is described in Legendre *et al* (1994) and Legendre & Legendre (1998). In short, the upper half of a symmetric distance matrix is collapsed into a vector, and a normal regression is performed on these collapsed vectors. Probabilities are subsequently assigned to the fit of the model and the beta coefficients, by randomizing the order of the values of the dependent variable while preserving its distance matrix structure, and calculating the regression again, and this process is repeated as many times as required for the level of significance against which the regression must be tested. The test statistic of the original data is compared to the frequency distribution of the test statistics of the permutations, and the ratio of values below and above this value gives a value for the probability of the regression model. In other words, the significance of the regressions is tested using a randomization test.

Randomization tests

Randomization tests, or permutation tests, are powerful methods to test the significance of a statistic with unknown distribution (Manly 1991, Good 1994). These tests are a good way to assign a probability to test statistics dealing with spatial epiphyte data. As long as all possible locations of the epiphyte are known within the sampled area, the censused locations of epiphytes can be tested against a random assignment of epiphytes on the various locations (VanDunné 2001). The test itself consists of the repeated calculation of a test statistic on random permutations of the original data. The outcome of these calculations is ordered so that the position of the test statistic calculated for the original data can be determined. Based on 9999 permutations, the 250 largest and 250 smallest values of calculated test statistics fall within the limits of a 5% significance level (two sided).

Since the test statistic used can be chosen at will, based on the data and hypothesis under consideration, permutation tests are flexible and can have a variety of applications. This flexibility is also the reason why one must have some basic programming skills to apply them. In general, high-end programming or scripting languages are more than sufficient for statistical modeling (Hilborn & Mangel 1997) and permutation tests in particular. There are many higher level languages available and each of these can be used to perform the calculations. Software is available on the internet for some specific applications (e.g. www.ecocam.com/software).

Conclusions

The spatial ecology of epiphytes and other canopy dwelling organisms is a fascinating subject. Much work still remains to be done. Placing epiphyte studies in their appropriate spatial context can only lead to a better theoretical framework of canopy ecology and increase the demand for statistical methods for three dimensional spatial analysis. Hopefully that demand will encourage statisticians to include chapters on three-dimensional analysis in their textbooks.

Contact details

Dr Hein JF Van Dunné & Dr Jan HD Wolf, Universiteit van Amsterdam Institute for Biodiversity and Ecosystem Dynamics (IBED), P.O. Box 94062, 1090 GB Amsterdam, The Netherlands.
e-mail: dunne@science.uva.nl, wolf@science.uva.nl

Literature cited

Bader M, VanDunné HJF & Stuiver HJ (2000). Epiphyte distribution in a secondary cloud forest vegetation; a case study of the application of GIS in epiphyte ecology. *Ecotropica* 6: 181–195

Bergstrom DM & Tweedie CE (1998). A conceptual model for integrative studies of epiphytes: Nitrogen utilisation, a case study. *Australian Journal of Botany* 46: 273–280

Borcard D, Legendre P & Drapeau P (1992). Partialling out the spatial component of ecological variation. *Ecology* 73: 1045–1055

Burrough PA (1987). Spatial aspects of ecological data. Pp. 213–251. in: Jongman RHG ter Braak CJF & van Tongeren OFR, eds. *Data analysis in community and landscape ecology*. Pudoc, Wageningen.

Crawley MJ (1997). The structure of plant communities. Pp. 475–531 in: Crawley MJ, ed. *Plant Ecology*. 2 ed. Blackwell Science, Oxford.

Cressie NAC (1993). *Statistics for spatial data*. Wiley, New York

Dale MRT (2000). *Spatial pattern analysis in plant ecology*. Cambridge University Press

Good P (1994). *Permutation tests, a practical guide to resampling methods for testing hypotheses.* Springer, New York

Gradstein SR, Hietz P, Lucking R, Lucking A, Sipman HJM, Vester HFM, Wolf JHD & Gardette E. (1996). How to sample the epiphytic diversity of tropical rain forests. *Ecotropica* 2: 59–72

Herben T, During HJ & Law R. (2000). Spatio-temporal patterns in grassland communities. Pp. 48–64 in: Dieckmann U, Law R and Metz JAJ, eds. *The geometry of ecological interactions: simplifying complexity*. Cambridge University Press, Cambridge

Hilborn R & Mangel M (1997). *The ecological detective; confronting models with data.* Princeton University Press, New Jersey

Hurlbert SH (1990). Spatial distribution of the montane unicorn. *Oikos* 58: 257–271.

Isaaks EH & Srivastava RM (1989). *Applied geostatistics*. Oxford University Press, New York

Johansson D (1974). Ecology of vascular epiphytes in West African rain forest. Acta Phytogeographica Suecica 59: 1–128

Koenig WD (1999). Spatial autocorrelation of ecological phenomena. Pp. 22–26 in: *Trends in Ecology and Evolution*

Legendre P (1993). Spatial autocorrelation: trouble or new paradigm? *Ecology* 74: 1659–1673

Legendre P, Lapointe FJ & Casgrain P (1994). Modeling brain evolution from behavior: a permutational regression approach. *Evolution* 48: 1487–1499

Legendre P & Legendre L (1998). *Numerical ecology*. Elsevier, Amsterdam

Lowman MD & Wittman PK (1996). Forest canopies: methods, hypotheses, and future directions. *Annual Review of Ecology and Systematics* 27: 55–81

Madison M (1979). Distribution of epiphytes in a rubber plantation in Sarawak. *Selbyana* 5: 207–213

Manly BFJ (1991). *Randomization and monte carlo methods in biology.* Chapman and Hall, New York

Nieder J & Zotz G (1998). Methods of analyzing the structure and dynamics of vascular epiphyte communities. *Ecotropica* 4: 33–39

Rossi ER, Mulla DJ, Journel AG & Franz EH (1992). Geostatistical tools for modelling and interpreting ecological spatial dependence. *Ecological Monographs* 62: 277–314

Shaw JD & Bergstrom DM (1997). A rapid assessment technique of vascular epiphyte diversity at forest and regional levels. *Selbyana* 18: 195–199

Van der Meer PJ (1997). Vegetation development in canopy gaps in a tropical rain forest in french Guiana. *Selbyana* 18: 38–50

VanDunné HJF (2001). Establishment and development of epiphytes in secondary neotropical forests. Dissertation, Universiteit van Amsterdam. 123 pp

VanDunné HJF (in press). Effects of the spatial distribution of trees, conspecific epiphytes and geomorphology on the distribution of epiphytic bromeliads in a secondary montane forest (Cordillera Central, Colombia). *Journal of Tropical Ecology*

Young LJ & Young JH (1998). *Statistical Ecology: a population perspective.* Kluwer, Boston

Toward a Sustainable use of Canopy Epiphytes as a Tool for Conservation

By
Jan HD Wolf

ESPITE consumer boycotts, educational programmes and international agreements such as the Convention on International Trade in Endangered Species (CITES), conservation efforts in tropical areas have often failed, apparently because they do not provide a direct economic benefit for local stakeholders. Consequently, they advocate an increase in land purchases as a means of financial compensation for conservation. As a complementary approach, some forest products could be harvested for commercial purposes without affecting the ecological integrity of the forest. Non timber forest products are most appropriate and recently it has been proposed to open up the potential of the canopy for this purpose, a concept known as canopy farming[1]. Vascular epiphytes are just one of the potential products in the canopy that may be removed with little impact. Because of their perceived beauty, orchids have been grown in cultivation for at least 2000 years, and the experience gained facilitates *in situ* and *ex situ* conservation efforts. Drawing on the commercial potential of orchids, a canopy farm for the production of seeds of low biomass, high value

A Mayan from Chiapas, Mexico with the inflorescence of *Tillandsia ponderosa,* found in the highlands. Guatemala exports 14.5 million *Tillandsia* plants each year. Photo: Antonio Turok.

Inflorescences of the epiphyte *Tillandsia eizii* are used in religious ceremonies by the Maya of Chiapas, Mexico. Seven species of Tillandsia are on the Appendix II of CITES. Photo: Fabián Ontiberos.

species was established in Costa Rica. The orchids are mounted on pieces of rooting substrate attached to nylon grids which are suspended in the forest canopy and can then be lowered to the forest floor for harvesting.

In this paper we pay particular attention to bromeliads, another widely appreciated group of ornamental epiphytes. On the American continent, bromeliads are often the most conspicuous epiphytes, because of their characteristic growth form and the large number of individuals which can exist in particular areas which have a well-defined dry season such as southern Mexico, Guatemala and El Salvador.

The history of bromeliad use and collection

The indigenous population of Mayan origin in the Highlands of Chiapas in the south-west of Mexico traditionally prefer to use bromeliads for ceremonial purposes and for the decoration of sacred sites. When asked why, people respond: 'because they appear in our dreams'. Interest in bromeliads in the Old World has for centuries been restricted to the pineapple. This terrestrial bromeliad has been in cultivation in The Netherlands since at least the end of the 17th century, judging from a painting, 'De Familie van Sybrand de Flines', by Jan Weenix (1642–1719; collection Amsterdam Historical Museum). While growing this rarity in the home was, presumably largely for decoration, it is only recently that the western world has in reality 'discovered' the potential of bromeliads for (home-) decoration. Many species in the genus *Tillandsia* are now regarded as a valuable cash crop in several countries. Information suggests that between 1993 and 1995, Guatemala exported 14.5 million *Tillandsia* plants annually. Rauh (1992) estimates that at least 75% of the plants in Guatemalan *Tillandsia* farms are collected from the wild. What remains unclear is how large a percentage of the plants gathered from the wild are of export quality, possibly not more than 10%. More important, than an extensive discussion about precise numbers is the observation that locally many populations are declining, casting doubts upon the sustainability of this harvesting practice. In an attempt to obtain clarity and to control export it was proposed in Kyoto (1992) to include all species in the genus *Tillandsia* in Appendix II of CITES. Finally, seven species of *Tillandsia* were added.

While local harvesting contributes to the decrease of plant populations, the most severe threat to epiphytes anywhere is probably the destruction of the forest they depend on. In addition to deforestation, selective logging of large trees and the destruction of the undergrowth especially threaten the least drought resilient species.

On a more positive note, the commercial activities associated with bromeliads show their potential to provide direct economic benefits for local communities. Sound management plans for sustainable exploitation are needed if bromeliads are to be used as a tool for overall forest conservation. Different approaches may be adopted for
• common species that attain high densities and
• low-density species, which are the majority.

Common species may be harvested from the wild if sustainability of yield can be guaranteed. Whereas species that are more rare need to be cultivated. If the harvesting of bromeliads is to be used as a tool for biodiversity conservation, it should also be compatible with socio-economic and ecological sustainability, topics which are not going to be addressed in this paper.

The harvesting of common bromeliads

Two approaches may be adopted to attain sustainability of yield. Demographic and genetic studies and modelling can be used in an attempt to establish the minimum viable population size and thus propose management interventions accordingly. This is known as population viability analysis. However reliable analysis, requires observations over a long period of time and has no universal value. For the pine-oak forests and bromeliads in the Highlands of Chiapas which are threatened at present, a more empirical approach to guarantee sustainability of yield has been suggested. Theirs is the first study that attempts to define guidelines for the sustainable use of wild populations and their method will be presented in detail in the following section.

Wolf & Konings (2001) start with the definition of the strictest criteria for sustainable harvesting. The thresholds are chosen arbitrarily but are based on generally accepted ecological principles. Applicable populations that are exploited should be monitored over time after which the stringent thresholds of the criteria can perhaps be lowered. They propose that harvesting should only be permitted from populations with:
1. a high population density
2. that are evenly distributed in space
3. and for which the reproductive potential will not be affected by the removal

Harvesting from a small population might negatively affect the local survival of a species, since small populations may experience reduced offspring fitness and a loss of genetic variability through inbreeding or genetic drift. In addition, they are considered to be more vulnerable to demographic and environmental stochasticity, and to natural catastrophes, (Young *et al* 1996). Wolf & Konings (2001) suggest utilising a minimum population density limit for exploitation, of 9,000 large rosettes/ha, which is a threshold ten times higher than an apparently stable *Tillandsia circinnata* population in Florida.

Second, they consider that populations evenly dispersed within a homogeneous habitat are at carrying capacity, for which the extinction risk is also smaller. Presumably for 3-D epiphytic populations at capacity, this implies that the abundance of epiphytes on trees of larger inhabitable size is nearly proportionally greater than the abundance on smaller trees. On the basis of a pilot study at the indigenous community 'La Florecilla', in forest stands along a disturbance gradient. They propose to employ the squared correlation coefficient (r^2), of a linear correlation of tree size against epiphyte abundance as an index of spatial homogeneity (ISH) of the population. The variable of tree size comprises the dbh and the number of branching points (forks in the branches with diameter >5 cm) of the tree, which are added after standardisation. Field tests showed that these two variables of the available space were estimated with much less error than other parameters such as tree height, crown volume, branch length or branch surface area. For the population of the common epiphyte *Tillandsia vicentina* at La Florecilla they suggest that harvesting should be limited to populations with an ISH >0.90 (p<0.001). The height of the threshold should not be regarded as universal and is likely to vary between epiphytic species and forest types. For different species and stands of forest in other places the ISH may be determined separately (in undisturbed stands).

The cultivation of locally abundant epiphytic bromeliad *Tillandsia fraseri* in Ecuador for export to the United States. Photo: Jan Wolf

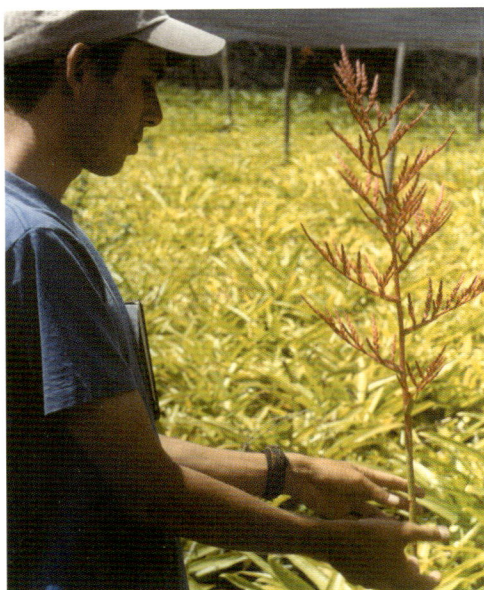

Third, to assure that the removal of rosettes does not affect the reproductive capacity of the population, Wolf & Konings (2001) propose to exploit only that part of the bromeliad population that grows in the lower stratum of the forest, including plants that have fallen on the forest floor. Population densities in the lower stratum of the forest are likely to be sink-populations that depend on resupply from the canopy, the environment to which they seem best adapted to survive. In contrast, canopy colonisation by wind dispersed seeds from lower strata seems unlikely and plants that grow near the forest floor are unlikely to play an essential role as providers of progeny for canopy populations of bromeliads.

Wolf & Konings (2001) also developed a user-friendly transect method that aims to identify populations of bromeliads that may be harvested with a sustainable yield. The method provides the required information on the density and spatial distribution of the population with minimal effort. No canopy access or specialised equipment is required and the data analysis is straightforward so that local communities may apply the method. As to bromeliad abundance, only rosettes larger than 20 cm are counted, which proved possible from the forest floor with the help of binoculars. No attempt was made to count true individuals (genets) of the clonal species, as the crowded growth of rosettes often made this impossible. A transect method was preferred over stratified sampling after field tests in this structurally heterogeneous forest showed that local stakeholders could not reliably distinguish and map forest stands that varied in the degree of disturbance and epiphyte abundance. Since stands of secondary forest are thus included, a high degree of variability between trees is expected because trees in recovering stands will not be occupied to capacity by bromeliads. The resulting distribution in a bivariate diagram of tree size and number of *T. vicentina* rosettes can be described as a 'factor-ceiling' distribution. In the calculation of the ISH only the host trees with a better than average occupation, i.e. the trees above the fitted line, the positive residuals, are considered.

In the 160 ha forest at La Florecilla, *T. vicentina* showed both a satisfactory average population density of c. 24,000 rosettes over 20 cm/ha on oaks and an ISH of 0.91. Less than 20% of the population occurred in the lower stratum of the forest, up to a height of 6 m. In compliance with the proposed prerequisites and taking into consideration some aspects of quality, Wolf & Konings estimate that it is possible to sustainably harvest c. 700 rosettes of *T. vicentina* /ha/yr from the understory and forest floor, in a 4-year rotation cycle. This amounts to an annual sustainable yield of 112,000 rosettes from the entire forest at La Florecilla. The plants that are removed from the lower stratum are being replaced naturally but the rate at which this happens and the influence of this removal on other forest organisms remains to be investigated.

The cultivation of relatively rare bromeliads

The cultivation of ornamental bromeliads has explosively increased over the last 10–15 years and from the Netherlands alone, hundreds of thousands of plants are exported annually, even to some tropical countries. However many canopy bromeliads that use the Crassulacean Acid Metabolic pathway (CAM), take too long to develop to be grown with commercial viability. One way bromeliad growers in the Netherlands overcome this constraint is through the development of new varieties, thus creating their own market. Energy restrictions in tropical areas are less severe, but a long-term outlook is still required in the cultivation of native CAM bromeliads. Commercial growing in greenhouses takes place only on a limited scale. Having said that in countries like Colombia, Ecuador and Guatemala, blooming businesses are now in operation.

   Faced with the high seedling mortality and slow growth exhibited by epiphytic bromeliads in the wild, *in vitro* culture techniques have been applied to propagate and enhance the early stages of development. *In vitro* propagation had earlier been used for the preservation of rare or difficult to propagate bromeliads such as *Tillandsia dyeriana*, a highly localised species endemic to Western Ecuador , and *Vriesea hieroglyphica*, *V. fosteriana*, and *Dyckia macedoi*, three endangered Brazilian bromeliads. Pickens *et al.* (in press) developed successful protocols for *Tillandsia eizii*, a heavily collected bromeliad endemic to the Highlands of Chiapas. The protocols enhanced seed germination, and produced plants with the characteristic tank morphology needed for survival in the wild. The growth rate of *Tillandsia eizii* juveniles in tissue culture, moreover, could be enhanced tenfold. Tissue cultured plants were successfully planted out in a greenhouse and have exhibited normal morphology. Whether juvenile plants that are thus given a head start can also be cultivated in the canopy of natural forests to increase their economic value merits investigation.

Conclusion

The potential of the canopy as a source for the sustainable extraction of vascular epiphytes and other non-timber products needs to be further investigated. If such products can be exploited sustainably without affecting the ecological integrity of the forest and for the benefit of local stake holders, then in many situations the canopy may hold the key for the conservation of tropical forests.

Note/Acknowledgement

1. I thank Oldeman RAA for permitting free use of the term: Canopy Farming©.

Contact details

Hein JF Van Dunné & Jan HD Wolf, Universiteit van Amsterdam, Institute for Biodiversity and Ecosystem Dynamics (IBED), P.O. Box 94062, 1090 GB Amsterdam, The Netherlands
e-mail: dunne@science.uva.nl, wolf@science.uva.nl

Table 1 Schematic presentation of the proposed method to identify populations of bromeliads that may be exploited with sustainability of yield (after Wolf & Konings 2001).

Reconnaissance
- *Select a group of species of host trees, having similar bark characteristics, that support dense bromeliad populations.*
- *Map the area.*
- *Lay out several parallel transects, covering the total area.*
- *Establish at least 35 random sampling points on these transects.*

Inventory
- *Select the four nearest trees to each point, one per quarter, with dbh >5 cm (point-centred quarter method).*
- *Record for each tree: Mean Distance (MD) to sampling point (cm), species, dbh and no. of branching points (diam.>5 cm).*
- *Record for the bromeliads: species, no. of rosettes >20 cm tall (in some cases smaller species may also be included), and the no. of rosettes in the lower forest stratum, i.e. up to a height of 6 m or approximately 1/3 of the total canopy height.*

Analysis
- *Calculate the Host Tree Density per ha (HTD); HTD = 10000/((MD/100)*(MD/100)).*
- *Calculate per bromeliad species the average occupation (O) per host tree; O = total no. of rosettes/number of trees.*
- *Calculate the Standard Error of the average Occupation (SEO); SEO = standard deviation/square root number of trees.*
- *Calculate per bromeliad species the average density per ha (BD); BD = HTD*O.*
- *Calculate the lower limit of the 95% confidence interval of the bromeliad density (LLBD); LLBD = HTD*(O-SEO*1.96)*
  IF LLBD <9.000 THEN STOP.
- *Standardise for all trees the dbh and the no. of branching points; standardised value = (value-mean)/standard deviation.*
- *Plot Tree Size (TS) against no. of rosettes; TS= sum of standardised dbh and no. of branching points.*
- *Fit a regression line through the bivariate diagram and select the data subset above the line (residuals >0)*
- *Calculate Index of Spatial Homogeneity (ISH); ISH = squared correlation coefficient between Tree Size and square root no. of rosettes.*
  IF ISH <0.90 THEN STOP.

Exploitation
- *Harvest bromeliads in the lower forest stratum up to 6 m, in a 4 year rotation cycle.*
- *Implement a monitoring program, applying the described method.*

Table 2 Method for the propagation of Tillandsia eizii in tissue culture

---

T. eizii seed are small (~2 mm) and have extensive coma hairs that aid in seed dispersal. The coma hairs were excised and the seeds were initially wetted by immersion in 1% NaOCl plus Tween 20 for 20 minutes, then rinsed with H_2O for five minutes. Seeds were disinfected using 70% ethanol for 2 minutes, followed by 2.6% NaOCl + Tween 20 for 40 minutes and rinsed in sterile water twice for 5 minutes each. Seeds were placed on a germination medium consisting of a modified Knudson's medium (KND) composed of Knudson's basal salts plus myo-inositol (0.1mg/l), nicotinic acid (5 mg/l), thiamine HCl (5 mg/l), glycine (4 mg/l), pryridoxine (5 mg/l) and gelled with 0.4% (w/v) Gelgro (ICN Biochem., Irvine, CA). Seed were germinated and grown in medium without sucrose until 12 weeks after initiation, then transferred to the same medium but with 2% sucrose. Media were adjusted to a pH of 5.5 prior to sterilisation and autoclaved at 120°C for 20 minutes at a PSI of 2.9 MPa. Seed were dispensed into 25x50 mm test tubes with 20 ml medium. After 8 weeks seedlings were transferred into 5.5 cm x 5.5 cm x 6 cm baby food jars containing 50 ml of KND medium. Cultures were maintained in a growth room at 25°C under a 16-hr photoperiod with illumination of 125 μmol/m2/s, and subcultured every 4 weeks to fresh medium.

Literature cited

Benzing DH (1978). The population dynamics of *Tillandsia circinnata* (Bromeliaceae): Cypress crown colonies in southern Florida. *Selbyana* 5, 256–263

Benzing DH (1990). *Vascular Epiphytes* Cambridge University Press, Cambridge, UK

Benzing DH (ed.) (2000). *Bromeliaceae: profile of an adaptive radiation*. Cambridge University Press, Cambridge

Breedlove DE & Laughlin RM (1993). The flora. In The flowering of man. A tzotzil botany of Zinacantán. (eds. Breedlove DE & Laughlin RM), *Smithson. Contr. Anthropol.* 35, Washington, D.C

Dimmitt M (2000). Endangered Bromeliaceae. In *Bromeliaceae: profile of an adaptive radiation* (ed D.H. Benzing), pp. 609–620. Cambridge University Press, Cambridge

Freese C (1996) *The commercial, consumptive use of wild species: managing it for the benefit of biodiversity.* World Wildlife Fund, Washington D.C.

Gullison RE, Rice RE & Blundell AG (2000). 'Marketing' species conservation. *Nature*, 404, 923–924

Hágsater E & Dumont V (eds.) (1996). *Orchids – status survey and conservation action plan*, pp 153. IUCN, Gland, Switzerland and Cambridge, UK.

Hietz P & Hietz-Seifert U (1995). Composition and ecology of vascular epiphyte communities along an altitudinal gradient in Central Veracruz, Mexico. *Journal of Vegetation Science* 6, 487–498

Holbrook NM (1991). Small plants in high places: the conservation and biology of epiphytes. *TREE* 6(10), 314

Knudson L (1946). A new nutrient solution for germination of orchid seed. *AOS Bulletin* 15, 214–217

Luther HE (1994). A guide to the species of *Tillandsia* regulated by Appendix II of CITES. *Selbyana*, 15, 112–131

Menges ES (2000). Population viability analyses in plants: challenges and opportunities. *TREE*, 15, 51–56

Mercier H & Kerbauy GB (1995). The importance of tissue culture technique for conservation of endangered Brazilian bromeliads from Atlantic rain forest canopy. *Selbyana* 16, 147–149

Neugebauer B, Oldeman RAA & Valverde P (1996). Key principles in ecological silviculture. In Fundamentals of organic agriculture. 11th IFOAM International Scientific Conference (ed Ostergaard TV), Vol. Vol.1, Copenhagen, Denmark.

Pickens KA, Affolter JM, Wetzstein HY, and Wolf JHD (in press). Enhanced seed germination and seedling growth of *Tillandsia eizii in vitro. Hortscience*

Rauh W (1992). Are Tillandsias endangered plants? *Selbyana* 13, 138–139

Rogers SE (1984). Micropropagation of *Tillandsia dyeriana. J. Bromeliad Soc.* 34, 111–113

Thomson JD, Weiblen G, Thomson BA, Alfaro S & Legendre S (1996) Untangling multiple factors in spatial distribution: Lilies, Gophers, and Rocks. *Ecology* 77 (6), 1698–1715

Véliz Pérez ME (1997). Los biotopos universitarios y la reintroducción de flora silvestre. In Proceedings Workshop *Rescate, Rehabilitación y Reintroducción de Vida silvestre*, pp. 63–70, Universidad de San Carlos de Guatemala, Guatemala

Verhoeven KJF & Beckers GJL (1999). Canopy farming: an innovative strategy for the sustainable use of rain forests. *Selbyana* 20(1), 191–193

Wolf JHD & Konings CJF (2001). Toward the sustainable harvesting of epiphytic bromeliads: a pilot study from the highlands of Chiapas, Mexico. *Biological Conservation* 101, 23–31

Young A, Boyle T & Brown T (1996). The population genetic consequences of habitat fragmentation in plants. *TREE* 11, 413–418.

6

Arthropods

Canopy Knockdown Techniques for Sampling Canopy Arthropod Fauna

By
Roger Kitching,
Yves Basset,
Claire Ozanne
and
Neville Winchester

SAMPLING the free-living arthropod fauna of the forest canopy using an insecticidal cloud has become the method of choice for general, rapid canopy collecting. Initially used for general inventory work the technique has become the preferred one for the ecological assessment of diversity in tree crowns (Basset 1997). The samples it generates are well suited to studies of trophic structure through guild analysis. In general, it is an ecological technique for comparative analysis where the ecosystematic response to a uniform or near-uniform sampling regime across sites and/or seasons is being assessed. Its usefulness in this context is substantially enhanced when used in conjunction with other sampling techniques both in the canopy and in adjacent forest components (e.g., Kitching *et al* 2000, 2001). It can also be used for collecting live specimens, especially for large and well-sclerotised insects (species which have hardened tissues in their exo-skeletons (Paarman & Stork, 1987).

Generally speaking knockdowns use a short-lived, quick acting pyrethrum (or pyrethroid) insecticide. In related but less generally used methods, local knockdowns often following bagging or otherwise enclosing segments of the tree crown, may be achieved using other agents such as carbon dioxide gas. The early history of the technique is described by Erwin (1990). One of the earliest users of the technique was Martin (1966) working in temperate forests. Roberts (1973) was the first to publish on the technique's application in tropical forests in his studies of Orthoptera in Costa Rica (although he used the much more powerful insecticide dichlorvos™). Gagné (1979) and Erwin (1982) modified these techniques to obtain quantitative samples of canopy faunas. Other key early references include Southwood *et al* (1982a, 1982b), Adis *et al* (1984), Stork (1987a, 1987b, 1988) and Stork & Hammond (1997).

Machinery and insecticides

Early use of canopy knockdown used a fogger – a machine based on a simple diesel motor delivering hot exhaust under pressure along a nozzle. An insecticide chamber is connected to the nozzle and once operating the machine delivers a cloud of diesel exhaust with very fine droplets of the insecticide. In perfect conditions this machine will produce a hot cloud of insecticide which will actually rise into the canopy beneath which the machine is being operated. The key words here are 'perfect conditions' – the slightest air movement can cause the cloud to move away from the segment of canopy you intend to sample (and beneath which you have placed your collectors). In the tropics dawn is often the time of least air movement in forest sites and has become the preferred time for sampling by fogging.

In North temperate forests later sampling times are preferred. Many workers now prefer to use a *mister* – a back-pack machine which runs using a two stroke petrol motor to deliver a spray of slightly larger insecticidal droplets into the canopy. This insecticide will not form a lighter-than-air cloud and so the insecticide must be delivered directly onto the target foliage. Ideally the machine is suspended on a rope from an emergent canopy tree such that the machine can be hauled into the midst of the canopy which is to be sampled. Although rather less sensitive than fogging this technique is still wind-sensitive and early mornings are again the preferred sampling time. Being less susceptible to drift than a thermal fog, the mist has a smaller radius of impact on the non-target areas of the canopy surrounding the sampling site.

There are many insecticides available and they all have different properties in terms of strength, persistence, vertebrate toxicity, miscibility and so forth. They may also require admixture with emulsifiers or carriers to increase their effectiveness. It is a matter of personal choice for the researcher but the trade-off may well be between knockdown efficiency, persistence and toxicity (requiring much greater safety precautions and the threat of non-target effects).

Many researchers prefer to use natural pyrethrum which is moderately efficient in knockdown effectiveness, has very short persistence properties and is relatively easily handled. Sometimes such natural pyrethrum is in short supply and it is always expensive (although a little goes a long way). Synthetic pyrethrums may be substituted and these come in many formulations. Some can be very toxic to things other than arthropods (particularly to fish) and difficult to use. Adis *et al* (1997) discusses many of the technical aspects of insecticide choice and formulation.

Site establishment and insecticide delivery

This account is based on that presented by Kitching *et al* (1992). The sampling of canopy over 12–15 m from the ground, demands that the mister or fogger is hoisted above the ground. A suitable, stout branch must be found at an appropriate height, over which a rope carrying a pulley though which has been threaded the rope attached to the mister or fogger can be hauled. The choice of an ideal branch will be determined by the goals of the study – in particular whether a tree-species specific sample or a mixed canopy sample is required. In practice such branches are in short supply and their presence often dictates exactly where the sampling will occur. In plot studies in which randomized samples are sought random coordinates generated for sampling locations need to be interpreted as 'the nearest suitable branch to location (x, y)' rather than 'the exact location (x, y)'. Lines can be placed over suitable branches using long- or cross-bows fitted to shoot fishing arrows, by sling-shots (catapults) that throw line-carrying lead weights (e.g., the convenient 'Big Shot'™), or even with line-throwing shot-guns. In all these cases the lightweight (usually fishing) line used for the initial shot is progressively replaced until a rope of sufficient weight is in place.

Sometimes the topography of the site or the height of the canopy allows the insecticide to be delivered directly by manual direction of the machine nozzle onto

Figure 1 Collecting funnels suspended beneath the canopy from a network of cords. Photo: Roger Kitching.

the canopy. In open forests such delivery may be made more efficient using scaffolds, 'cherry-pickers' or other elevated platforms. Canopy knockdown using insecticide is not recommended for use at fixed locations such as the crane sites because of the potential for non-target effects on parallel research activities.

In most rain forest situations a network of lightweight cord at about head height can be woven around the location where the sample will fall. A ten-by-ten metre area is often used. From this 'cats cradle' of cordage collecting funnels are suspended (Figure 1). The exact size or number will depend on the sampling intensity required. Twenty half-metre square funnels is a commonplace sampling intensity. Each collecting hoop is made of plasticised fabric with an elasticised cylinder or screw thread stitched into its base into which a plastic collecting vial can be inserted. This is usually set up with a little ethanol (or other fixative) in it. In some circumstances collecting hoops or vials may need drain holes with mesh over them to cope with excessive liquid in the event of rain. Schematics of hoop construction are provided in Kitching *et al* (2000) (available upon request from the author). If the actual size of the catch is to be maximised (for instance, for taxonomic purposes) then sheets may be spread on the ground outside the areas with funnels to provide access to as higher proportion as possible of the affected arthropods.

The actual spraying event usually comprises delivery of the insecticide into the canopy for up to 5 to 6 minutes. In general, times for misting will be less than for fogging and, depending on the intended size of the area to be sampled, may be as short as half a minute. Usually the machine is started on the ground and then rapidly hauled into the canopy by rope (for misters an angle-iron frame is a good substitute for the human-back for which they are designed). Some workers have used radio-controlled valves so that the actual dispensing of the insecticide does not begin until after the machine is in the canopy itself (Adis *et al* 1998). For tall trees (>30 m), the best fogs are achieved by climbing with single rope techniques into the tree crown, with the fogger suspended on a second line (e.g. Basset 1995). Sometimes it is useful to repeat the spraying event from the ground to collect understorey insects. If an upper and an understorey sample is to be collected and analysed separately then at least a day should separate the samples and the understorey should be sampled first. Although repeat sampling of the same site has been shown to produce rich samples even within 24 hours it is probably good practice not to re-spray the same forest components in any one season.

After spraying arthropods are collected by lightly brushing the funnels so that individuals are swept into the collecting vials. Specimens should be allowed to fall onto the trays for a standardised period of time, usually 2 to 4 hours, before removing and capping the vials.

Experimental design considerations

There are a number of considerations to be taken into account in designing surveys using canopy knockdown techniques. The following points are important.

- the sampling event (replicate) is the whole spraying event and catches from each of the collecting funnels should be combined before analysis. If different numbers of funnels have been used then counts can be converted to catches per square metre before analysis. Individual funnel catches are *not* replicates and cannot be used as such (in exceptional circumstances the catches of funnels close to a tree trunk (say) may be compared with those beneath the crown of a tree but the sampling event should be designed specifically to answer the underlying methodological questions and not simply an *a posteriori* analysis.
- a minimum of three samples must be made in any one location so that variances can be calculated. Probably five to ten samples are to be preferred. Spacing of such replicate sampling events should be randomized and, experience suggests, a minimum of 30 to 40 metres apart to minimise interaction effects. We await an experimental evaluation of the effect of inter-sampling distance.
- the use of sampling to compare estimates of species richness across sites presents some statistical problems and techniques such as rarefaction(e.g. Magurran 1988) will be required to compensate for different sizes of catch.
- in comparative assessments (across forests for example) ideally all samples should be collected simultaneously. This is virtually never possible for logistical reasons. Accordingly efforts must be taken to ensure comparability by sampling in the same season and at the same time of day.

General Do's and Don'ts
Do's

- begin canopy sampling sufficiently early in a field trip so that several unusable mornings (with winds that make such sampling impossible) can be accommodated.
- carry tools and spare parts for the fogger or mister – these machines require constant cleaning, maintenance and general 'tweaking' if they are to work efficiently.
- use approved goggles, aspirators and other safety equipment – ensure that filter discs are changed regularly according to the codes of practice set out for the insecticide – the rope haulers and the spray director should be so equipped.
- store insecticide and fuel safely both during transportation and in the field
- check that all the suspended funnels contain collecting vials with ethanol (or other fixative) before beginning the spraying event.
- check all collecting vials contain appropriate labels before beginning the sampling event – then check it a second time.
- hang collecting funnels from the 'cats cradle' using simple angling swivel clips secured, if necessary, by clothes pegs.

- ensure any insecticidal spillage on exposed skin or clothing is dealt with immediately using copious quantities of water.
- be aware of safety procedures specific to your particular insecticide.

Don't
- forget these machines need fuel as well as insecticide
- encourage on-lookers – for safety reasons
- haul the machine into the canopy before getting the heaviest member of the party to swing on the rope – a fuel- and insecticide-laden machine is a very awkward missile should the supporting branch give way (and it can!)
- let the fogger nozzle point downwards – diesel trickling into the hot nozzle makes for a very efficient flame-thrower (this is not a problem with misters)
- hang collecting funnels on the 'cats-cradle' the night before the sampling – they are very efficient rain collectors – far better to suspend them just before the sampling event itself
- use pyrethrum or pyrethroids near running or standing water – the side effects on the fish fauna may not be appreciated by others (or the fish!).

Comparison with other techniques
Knockdown compares very favourably with other canopy sampling techniques for arthropods (Ozanne, In press). The proportion of the fauna collected is higher than for beating and sweeping (Lowman *et al* 1996) and knockdown samples are highly representative of the structure of the canopy communities (Southwood *et al* 1982b). Although knockdown does underestimate sessile insects (Majer & Recher 1988) and provides a less effective method of estimating biomass than branch clipping, it is more efficient at sampling mobile and cryptic insects and collects a greater range of species than branch clipping (Blanton 1990).

Contact details
Professor Roger L Kitching, Australian School of Environmental Studies, Griffith University, Brisbane, Qld 411, Australia. Email: r.kitching@mailbox.gu.edu.au

Dr Yves Basset, Smithsonian Tropical Research Institute, Apartado 2072, Balboa, Ancon, Panamá City, Republica de Panamá. Email: BASSETY@tivoli.si.edu

Dr Claire Ozanne, School of Life Sciences, Roehampton University of Surrey, West Hill, London SW15 3SN, UK. Email: c.ozanne@roehampton.ac.uk

Dr Neville Winchester, Biology Department, University of Victoria, P. O. Box 3020, Victoria, British Columbia V8W 3N5, Canada. Email: tundrast@UVVM.UVIC.CA

Literature cited
Adis J, Lubin YD & Montgomery GG (1984). Arthropods from the canopy of inundated and Terra firme forest near Manaus, Brazil, with critical consideration on the Pyrethrum-fogging technique. *Stud. Neotrop. Faun. Env.*, 19, 223–236.

Adis J, Paarmann W, da Fonseca CRV & Rafael JA (1997). Knockdown efficiency of natural pyrethrum and survival rate of living arthropods obtained by canopy fogging in central Amazonia. In: (eds. Stork NE, Adis J & Didham RK) *Canopy Arthropods*, Chapman and Hall, London, pp. 67–81.

Adis J, Basset Y, Floren A, Hammond PM & Linsenmair KE (1998a) Canopy fogging of an overstorey tree – recommendations for standardization. *Ecotropica* 4, 93–97.

Basset Y (1995). Arthropod predator-prey ratios on vegetation at Wau, Papua New Guinea. *Science in New Guinea*, 21, 103–112.

Basset Y (1997). In: (eds. Stork NE, Adis J & Didham RK) *Canopy Arthropods*, Chapman and Hall, London, pp. 00–00.

Blanton CM (1990). Canopy arthropod sampling: a comparison of collapsible bag and fogging methods. *Journal of Agricultural Entomology* 7, 41–50.

Erwin TL (1982). Tropical forests: their richness in Coleoptera and other arthropod species. *Coleopterists' Bulletin* 36, 74–75.

Erwin TL (1990). Canopy arthropod biodiversity: a chronology of sampling techniques and results. *Revista Peruana de Entomologia* 32, 71–77.

Gagné WC (1979). Canopy-associated arthropods in *Acacia kea* and *Metrosideros* tree communities along an altitudinal transect on Hawaii. *Pacific insects* 21, 56–82.

Kitching RL, Bergelson J, Lowman MD, McIntyre S & Caarruthers G (1993).The biodiversity of arthropods in Australian rain forest canopies: introduction, methods, study sites and ordinal results. *Australian Journal of Ecology* 18, 181–191.

Kitching RL, Vickerman G, Laidlaw M & Hurley K (2000). *The Comparative Assessment of Arthropod and Tree Biodiversity in Ol-World Forests: the Rainforest CRC/EARTHWATCH Protocol Manual.* Technical Report, Cooperative Research Centre for Tropical Rainforest Ecology and Management, Cairns (70pp).

Kitching RL, Li DQ. & Stork NE (2001). Assessing biodiversity 'sampling packages': how similar are arthropod assemblages in different tropical rainforests? *Biodiversity and Conservation* 10, 793–813.

Lowman MD, Kitching RL & Carruthers G (1996). Arthropod sampling in Australian subtropical rain forests: How accurate are some of the more common techniques? *Selbyana* 17, 36–42

Majer JD & Recher HF (1988). nvertebrate communities on Western Australian eucalypts: A comparison of branch clipping and chemical knockdown procedures. *Australian Journal of Ecology* 13, 269–278

Martin JL (1966).The insect ecology of red pine plantations in central Ontario. IV. The crown fauna. *Canadian Entomologist* 98, 10–27.

Ozanne CMP (In press).Techniques and methods for sampling canopy insects. In: (ed. S. Leather S.) *Insect Sampling*, Blackwell Scientific, Oxford.

Paarmann W & Stork NE (1987).Canopy fogging: a method of obtaining living insects for investigations of life-history strategies. *Journal of Natural History* 21, 563–566.

Roberts HR (1973).Arboreal Orthoptera in the rain forests of Costa Rica collected with insecticide: as report on grasshoppers (Acrididae) including new species. *Proceedings of the Academy of Natural Sciences, Philadelphia* 125, 46–66.

Southwood TRE, Moran VC & Kennedy CEJ (1982a). The richness, abundance and biomass of arthropod communities of trees. *Journal of Animal Ecology* 51, 635–650.

Southwood TRE, Moran VC & Kennedy CEJ (1982b). The assessment of arboreal insect fauna: comparisons of knockdown sampling and faunal lists. *Ecological Entomology* 7, 331–340.

Stork NE (1987a). Guild structure of arthropods from Bornean rain forest trees. *Ecological Entomology* 12, 69–80.

Stork NE (1987b). Arthropod similarity of Bornean rain forest trees. *Ecological Entomology* 12, 219–226.

Stork NE (1988). Insect diversity: facts , fiction and speculation. *Biological Journal of the Linnean Society* 35, 321–337. 69–80.

Stork NE & Hammond PM (1997). Sampling arthropods from tree-crowns by fogging with knockdown insecticides: lessons from studies of oak tree beetle assemblages in Richmond Park (UK). In: (eds. Stork NE, Adis J and Didham R) *Canopy Arthropods*, Chapman and Hall, London, 1–26.

Canopy Micro-arthropod Diversity: Suspended Soil Exploration

By
Neville
Winchester

THE investigation of temperate and tropical ancient rainforests and the arthropod species assemblages associated with suspended soils in them, is a line of research and development of sampling protocols that has been ongoing since 1992. Funding has come from a variety of sources but is currently supported by the Natural Sciences and Engineering Research Council of Canada (NSERC) research grant to the author.

In general, canopies of rainforests contain a large percentage of the species present in forest systems and the most speciose group is the arthropods (see Stork *et al* 1997). Canopies of natural forests in temperate (Schowalter & Ganio 1998; Winchester 1997) and tropical regions (Erwin 1983, Stork 1987, Basset 1997, Didham 1997, Kitching *et al* 1997) contain largely undescribed and little understood assemblages of arthropods that have greatly expanded estimates of the total number of insect/arthropod species. However, it is not widely appreciated that microarthropods, primarily the Acari, are a dominant component of this fauna (Walter & Behan-Pelletier 1999). In my research I quickly realized that most arboreal arthropod species are small, especially mites, and I had better align myself with some leading taxonomists! The Acari are the most speciose and abundant arachnid group in arboreal microhabitats (Winchester 1997, Behan-Pelletier & Winchester 1998, Fagan 1999). Most canopy sampling methods for estimating total diversity (e.g. insecticide fogging) are not efficient for the collection of these organisms (Walter 1995; Walter *et al* 1998). The range of microhabitats available for colonization is also large (Winchester 1997). There are far too many microhabitats to give any researcher the satisfaction of implementing a holistic sample design that captures the very essence of diversity (only in our dreams) and thus it is thought that current estimates of the diversity of mites living in rainforest canopies are an underestimate (Walter & O'Dowd 1995, Fagan 1999). Within the Acari, oribatid mites (Oribatida) typically form the most diverse taxonomic assemblage (Wallwork 1983) and have the highest relative abundance of any of the suspended-soil microarthropods. They are also rather 'cute', at least under the microscope!

Traditionally, the forest canopy has been viewed as a substrate source for decomposition processes that occur, after litter falls to the forest floor (Fagan 1999). However, decomposition also occurs within the canopy and provides a substrate that supports a diverse invertebrate community. Many oribatid species associated with these substrates show little evidence of trophic specialization and species ostensibly co-exist (Anderson 1978). Variation in feeding habits has been used to classify species into six main feeding types (Schuster 1956; Luxton 1972). For example, certain oribatid species are associated with lichens or corticolous habitats (i.e. growing on or in tree bark), and it is now clear that distinct species assemblages are present in 'suspended soils' (*sensu* Wallwork 1976). With rich accumulations of debris and litter, temperate and tropical forest canopies provide the habitat template for an incredibly diverse range of microarthropods, particularly within the Acari and more specifically, within the oribatid mites.

Site selection
Large scale
Arthropod communities associated with suspended soils/epiphytes in the canopies of ancient forests are the central component of our research program. Therefore, we will go wherever ancient rainforest exists (unexplored is best) in our quest to investigate arboreal species assemblages. In particular this research is geared towards documenting species richness and examining factors that shape the distribution, diversity, abundance and community composition of resident arthropods in high-canopy ecosystems. For example, our recent work on canopy arthropods in the ancient coastal forests of Vancouver Island, in British Columbia, Canada indicates that numerous species, many of which are new to science exist in these forests. Many of them are specific to particular microhabitats (e.g. suspended soils/epiphytes). Similarly such species distribution patterns (although involving different species) seem to exist in other ecosystems, as evidenced by our work in Gabon, Malaysia and Costa Rica (it is exciting, to discover similar trends across widely separated forests).

In addition, statistical rigor is feasible since we use a standardized, previously tested and validated program to sample suspended soil arthropods. For these reasons, future investigations will be able to answer basic questions about the ecology of forest canopy arthropods. They will also be able to determine the key factors needed to empirically address conservation questions within and across ecological gradients not only at site-specific but also across wide geographic scales.

Small scale
In general, the single rope technique (SRT) is the access method that we use on our canopy expeditions (Figure 1). Basically, whatever fits into a backpack and can be easily carried to areas with low to no access, is the rule of thumb. Access techniques with specifics on SRT are adequately described in the literature (Barker and Sutton 1997). Tree selection takes into account recommendations made from the author's 1992–1999 canopy studies. Trees with suspended soil located in a continuous manner along branches at different heights in the canopy are ideal. Naturally, forest type, tree species and architecture dictate whether debris accumulations occur. Many trees are devoid of large amounts of suspended soil or have a patchy distribution of 'soil pockets' which presents some difficulty in terms of sampling procedures and also necessitates some changes in sample design. What is common to most trees that we have sampled, regardless of geographic location, is suspended soil accumulations layered with epiphytes and mosses. If deep suspended soil accumulations are absent then lichens are usually abundant and provide microhabitats for a number of species and a switch is made to sample lichens in lieu of suspended soils.

Figure 1 Suspended soil sampling in the Belum forest, Malaysia. Photo: Kevin Jordan

Experimental design and suspended soil sampling

There are a number of factors to consider when sampling suspended soil accumulations. The following protocols are typical for a sampling program collecting microarthropods associated with suspended soils.

- the main factors are tree species, tree height, horizontal position and epiphyte association.
- samples of suspended soil cores 3 cm diameter x 5 cm deep are collected in each tree using a simple and inexpensive garden bulb-planter. A total of 3 replicates is collected from each of 3 heights and 3 distances along a randomly chosen branch, making 27 samples in total from each tree. Generally, I like to sample a minimum of 5 trees per species, which means that there would be a minimum of 135 samples collected for each species.
  NOTE: In many trees only 2 heights (lower and upper canopy) can be delineated.
- replicate soil/litter core samples are also taken from the ground at each sample tree. I usually take 5 cores at the base of each tree sampled. This allows for a species comparison of ground versus canopy habitats (i.e. identification of arboreal-specific species).
- digital pictures could be recorded at each sample position within the canopy and would provide a method to deal with the identification of epiphyte diversity and distribution. Other variables such as mean suspended soil depth, moisture content, and the distribution and identification of epiphytes and mosses should be recorded. Collection of this information is dependent on the research objectives and varies with each project.

Sample processing in the field

The microarthopods associated with canopy samples need to be extracted within 72 hours. I have found the following protocols useful:

- Download the digital pictures and associate them with the collected samples.
- Tree species, epiphyte composition, height and canopy position should be associated with each sample.
- Arthropods are removed from each core by submersing the contents in a saturated saline solution in plastic bags for 48 hours. The samples are first washed through a 120 micron sieve into the saline solution. Contents of the saturated saline solution are then sieved, emptied into a whirl-pak bag and preserved in 75% EtOH. Height from ground and distance from the trunk are recorded for each sample. Additional details for the sampling and extraction of micorarthropods can be obtained by referring to: www.cciw.ca/eman-temp/reports/publications/sage/sage 10.htm

Data analyses

Individuals extracted from suspended soils can be analysed in a sequence, formalized into a series of seven steps by Winchester (1999). Each successive step adds complexity (and cost) to the analysis, but at the same time increases sensitivity. The first three steps require arthropod sorting to ordinal or family level, but give only coarse-grained resolution. Subsequent steps require identification to species level for designated taxa, but provide fine-grained sensitivity to answer detailed questions. Useful information about trophic interactions can sometimes be obtained by analysis of guilds at the family level, but resolution of pivotal questions using species-level identification is generally desired. An important message – a major requirement is collaboration with systematists in order to ensure that correct identifications are provided.

The seven-step package of analytical categories includes (for detailed explanations see Winchester 1999):

1. Total arthropod diversity (richness and evenness)
2. Total arthropod biomass and energy cycling
3. Trophic group analysis for all arthropods
4. Species richness and estimated total regional biodiversity
5. Changes in trophic group proportions
6. Species complementarity
7. Relating species composition to environmental gradients

Future direction

The information presented in this paper is intended to harmonize the collection and analysis of suspended soil arthropods by using a rapid, repeatable sampling methodology and analyses. The ultimate goal of this research is to document species richness and examine common factors that shape the distribution, diversity, abundance and community composition of resident arthropods in high-canopy ecosystems (see Forest Canopy Research Planning Workshop: Final Report, 1999, Oxford, UK). Not only is this information essential for determining global biodiversity patterns, but it will also provide linkages among study sites of direct relevance to global forest biodiversity issues.

Contact details

Dr. Neville N. Winchester, Biology Department, University of Victoria, Victoria, British Columbia, Canada. V8W 3N5. Phone: (250) 721–7099, Fax: (250) 721–7120. Email: tundrast@uvvm.uvic.ca

Literature cited

Anderson JM (1978). Inter- and intra-habitat relationships between woodland Cryptostigmata species diversity and the diversity of the soil and litter microhabitats. *Oecologia* 32: 341–348.

Barker MG & Sutton SL (1997). Low-tech methods for forest canopy access. *Biotropica* 29: 243–247.

Basset Y (1997). Species abundance and body size relationships in insect herbivores associated with New Guinea forest trees, with particular reference to insect host-specificity. Pp. 237–264 In Stork NE, Adis JA & Didham RK (eds.), *Canopy Arthropods*. Chapman and Hall, London.

Behan-Pelletier VM & Winchester NN (1998). Arboreal oribatid mite diversity: colonizing the canopy. *Applied Soil Ecology* 9: 45–51.

Didham RK (1997). Dipteran tree-crown assemblages in a diverse southern temperate rainforest. Pp. 320–343 In. Stork NE, Adis JA & Didham RK (eds.), *Canopy Arthropods*. Chapman and Hall, London.

Erwin TL (1983). Tropical forest canopies, the last biotic frontier. *Bullentin Entomological Society America* 29: 14–19.

Fagan LL (1999). *Arthropod colonization of needle litter on the ground and in the canopy of montane Abies amabilis trees on Vancouver Island, British Columbia.* MSc. thesis, University of Victoria, Victoria, British Columbia. Pp. 172.

Kitching RL, Mitchell H, Morse G & Thebaud C (1997). Determinants of species richness in assemblages of canopy arthropods in rainforests. Pp. 131–150 In Stork NE, Adis JA & Didham RK (eds.), *Canopy Arthropods*. Chapman and Hall, London.

Luxton M (1972). Studies on the oribatid mites of a Danish beechwood soil. 1. Nutritional Biology. *Pedobiologia* 12: 434–463.

Schowalter TD & Ganio LM (1998). Vertical and seasonal variation in canopy arthropod communities in an old-growth conifer forest in southwestern Washington, USA. *Bulletin of Entomological Research* 88: 633–640.

Schuster R (1956). Der Anteil der Oribatiden an den Zersetzungsvorgängen im Boden. *Zeitschrift für Morphologie und Ökologie der Tiere* 45: 1–33.

Stork NE (1987). Guild structure of arthropods from Bornean rain forest trees. *Ecological Entomology*. 12: 61–80.

Stork NE, Adis JA & Didham RK (eds.) 1997. *Canopy Arthropods*. Chapman and Hall, London.

Wallwork JA (1976). *The Distribution and Diversity of Soil Fauna*. Academic Press, London.

Wallwork JA (1983). Oribatids in forest ecosystems. *Annual Review of Entomology* 28: 109–130.

Walter DE (1995). Dancing on the head of a pin: mites in the rainforest canopy. *Records of the Western Australian Museum Supplement* 52: 49–53.

Walter DE & Behan-Pelletier VM (1999). Mites in forest canopies: filling the size distribution shortfall? *Annual Review of Entomology* 44: 1–19.

Walter DE, Seeman O, Rogers D & Kitching RL (1998). Mites in the mist: how unique is a rainforest canopy knockdown fauna? *Australian Journal of Ecology* 23: 501–508.

Walter DE & O'Dowd DJ (1995). Life on the forest phylloplane: hairs, little houses and myriad mites. Pp. 325–349 In Lowman MD &. Nadkarni NM (eds.), *Forest Canopies*. Academic Press, London.

Winchester NN (1997). *Conservation of Biodiversity: Guilds, Microhabitat Use and Dispersal of Canopy Arthropods in the Ancient Sitka Spruce Forests of the Carmanah Valley, Vancouver Island, British Columbia*. PhD thesis, University of Victoria, Victoria, British Columbia.

Winchester NN (1999). Identification of potential monitored elements and sampling protocols for terrestrial arthropods. Pp. 227–314 In Farr DD, Franklin SE, Dixon EE, Scrimgoeur G, Kendall S, Lee P, Hanus S, Winchester NN & Shank CC. Monitoring forest biodiversity in Alberta: program framework. *Alberta Forest Biodiversity Monitoring Program Technical Report 3* (draft) (http://fmf.ab.ca/bm.winchest.htm).

Collections and Sorting of Insect Frass

By
Eric John Olson

FOR the past 4 years I have been directing a research program aimed at tracking the seasonal dynamics of insect herbivory in a tropical dry forest. Our frass collections allow us to estimate whole-forest herbivory levels in several highly deciduous (and one primarily evergreen) dry forest tracts in the Santa Rosa Sector of the Guanacaste Conservation Area, northwestern Costa Rica.

Choosing a location

I chose Santa Rosa in part because the life histories of herbivorous insects are being well documented there, through long-term research by Dr. Daniel Janzen (University of Pennsylvania). Janzen's massive caterpillar rearing effort has revealed a wide diversity of life-history patterns, but nearly all participating species are active during the first third of the wet season. Most univoltine (producing only brood each year) species are only active at this time, and then either migrate away from the dry forest as adults, or remain in their pupae until the following year. My research aims to quantify the activity of the whole herbivore community on a twice-monthly basis to provide the context within which the life-histories of individual species can be viewed. We also aim to test Janzen's impressions that:

1. The early peak in caterpillar activity lags somewhat behind leaf expansion in a typical year
2. The peak occurs when many natural enemy populations are at a low ebb.

I also chose Santa Rosa as a location because I am familiar with it and this makes leading volunteers around from Earthwatch Institute, who assist in my work, much easier. In addition, the new Tropical Dry Forest Research Center which has been established in Santa Rosa has comfortable dormitories, labs and storage spaces, a small but useful library, and a conference room for evening lectures. The Costa Rican government also continues to welcome scientists, charging quite modest fees for research and specimen-export permits.

Equipment and installation

The methods we use are inexpensive, forest- and volunteer-friendly, and are easily replicated. There are a variety of ways to track canopy herbivore activity, but frass collecting best suits my goals and this particular site.

Frass work hinges on Olson's Law of Excretion, which bluntly states that 'If Something Eats it Must Eventually Poop'. We catch the excreta falling from the canopy on clear plastic sheets 85 cm long on each side, suspended just off the ground in a bowl shape so that the very center of the trap usually rests lightly on the ground. A smooth clean stone set in the center of the trap helps define this bowl, and reduces the frequency of wind-throw (and when the stone is tossed out that tells you wind or some passing animal has thrown a trap). Clear plastic is best because you can see snakes through it, though I instruct volunteers to peek

underneath whenever they kneel down to work with a trap, or before sticking their hands under one. We reinforce each corner with two applications of duct tape, then bang a small hole through the duct tape with a hole punch. For poles we cut strong galvanized wire, about 5 mm in diameter, into approximately 60 cm lengths. We bend one end into an S, on which we hook the frass-trap corners. Don't close the S down tight, since you'll want to slip the plastic sheet off, to shake out rain and accumulations of leaves and other crud when setting it up for a collection.

We collect from 240 traps each month, these are set up in eight grids laid out six rows wide (A to F) and five traps deep, so 30 traps total. There's 10 meters between rows and traps, and therefore the total area of forest floor occupied by a grid is 2000 m^2. We select forest tracts based on our understanding of local forest history, our aim being to conduct sampling in a range of different aged forest. We identified four forest tracts each presumed to have a unique history of disturbance, and within these we set up two grids using quasi-random placement to select their exact locations. That is, we chose our sites in the lab looking at a map, not in the forest looking at the trees, but we kept close to trails so that volunteer teams could readily find their way to and from grids. We also had to avoid steep ravines and existing study sites, and we didn't want the two sites in any given tract side by side. The result was decidedly quasi-random, and when we went out to the chosen areas we were not entirely happy with our choices. The forest just didn't 'seem' representative in every case. Of course that feeling is expected whenever sampling happens randomly, and of course we went ahead and used those pre-determined sites and haven't looked back.

Grids were carefully laid out using compass and tape to ensure accuracy. A distance between traps of 10 meters was chosen because at that distance one is usually no longer underneath the canopy of trees which are contributing frass to the nearest neighbouring traps. Such spacing helps increase the number of plant species, and thus of herbivore species, contributing frass to the traps, which is surely a reasonable aim in a community study of this sort.

When we choose to collect frass from a particular grid, we start by shaking out rainwater and debris from each trap. After getting every bit of dirt and moisture off the 30 sheets we perform a brief frass-dance, asking for 24 hours free of rain. This rarely works but fortunately there are often rain-free days in the dry forest habitat, even in the wettest months. I would guess frass-trapping would be difficult in most tropical or temperate rain forest sites due to the levels of rainfall. If it doesn't rain overnight we hustle back the next morning with 30 petri dishes, 5 cm in diameter, each with snug-fitting lids and containing a slip of paper listing site, date, and grid coordinates (A1, F5, etc.). At each trap we scoop up all the fine stuff into the petris: frass plus flower scales, orts, fallen stipules, seeds and so on. Back at the lab we dry the collection at about 40° C, then move the trays of petris to the lab.

Sorting happens under good illumination, we all wear jeweler's magnifiers and work with forceps and fine paint brushes. I and my field assistant spend a lot of time fielding questions at this stage, of the 'what the heck is THIS?' variety, but volunteers rapidly gain confidence and its really not rocket science to do this kind of work. We keep all the 'waste', and periodically I or an experienced assistant sits down with a fully sorted collection to double check it. Such a check usually results in moving a few frass pellets from the waste back to the collection, and some frass-mimicking seeds or dirt clumps in the other direction. Sorted collections are dried again at 40° C, then weighed. We use a Denver Instrument digital balance that records to 0.001 gram and usually behaves itself very well, and we frequently double-check it with small weights of known mass. We avoid combining the frass from separate dishes, doing so only when a collection is particularly skimpy (as often happens during April and other dry season months).

Canopy fogging is an alternative technique that might work, although I suspect a lot of fogged caterpillars die and rot up in the canopy, hanging from their remarkably tenacious crochets. Not all the frass falls to the ground either of course, and leaf cutter ants make no frass – so beware of what you claim for this method. The main problems with fogging are its cost and its political incorrectness in the context of a National Park, where destructive sampling can be viewed as unacceptable. It remains appropriate when vouchers are needed, and for single-tree species studies, but here the focus is on the activity of the forest community as a whole. Other possibilities for following the fates of individual canopy leaves are jumaring up (using ropes) or working from cranes.

Research projects

The core data of this frass research are of course snapshots of dry weight grams of frass per hour, over all months of the wet season. In addition we do monthly canopy photography and we also keep track of rainfall, through the Guanacaste Conservation Area weather station. We look forward to studying frassfall during an El Niño event, which brings intense droughts to Guanacaste. At the time of writing (2001) we are between El Niño and La Niña regimes, and we hope that means we're getting a few good years of 'typical' frassfall. Of course a typical climate regime can still be punctuated by extreme weather events, and we've definitely had one: Hurricane Mitch in October of 1998. The following year was eerily quiet in terms of herbivory, and most strikingly there was no frassfall peak at all. The forest looked gorgeous, showing very little foliar damage but that's just my impression. The frassfall data are clear though; the curve truly was flat throughout 1998. A peak reappeared in 2000, but quite moderate in size, and a stronger peak appeared in 2001. Perhaps several consecutive years lacking such sudden large rain events favor high survivorship of soil-dwelling pupae, setting the stage for major peaks in herbivory. Many additional years of collecting are in order, so stay tuned.

After we weigh and record the collection from a given petri dish, the frass is saved in small bags. This frass collection will then be used for at least two purposes:

1. The identification of major classes of insects by their excreta, and we can already report that dry season frass fall is primarily orthopteran; the lepidoptera don't kick in until the rains begin. I don't know what to expect with late wet-season frass, and I have not yet looked closely at it.
2. Running nitrogen and phosphorus studies to look at how frass contributes to nutrient cycles. Frass is about 2% N, about the same as green leaves, and there's an early wet-season peak of the stuff. This should result in a burst of relatively mobile N (maybe too mobile), falling from the canopy (many legumes in tropical dry forest) and working its way through the rooting zone of all the plants.

One reason why frass studies are a very good method of following the fate of leaves is because personal anecdotal evidence suggests that much of the damage to dry

forest foliage happens amazingly fast, probably too fast for leaf-fate studies to keep pace. For example, in 2001 we did hemispherical canopy photography and at the same time we were collecting frass, and found that the peak herbivory onslaught occurred during leaf emergence, as the canopy was in the process of closing. This means many leaves were being consumed whole just as they were expanding. The canopy was almost fully closed in just 3 weeks in spite of all the caterpillar damage, although a few species of Santa Rosa tree are chronically susceptible to total defoliations at this time of year. I find it difficult to imagine how leaf-tracking studies could handle such a fierce burst of damage. Though, which of the methods is best, depends on the level of the study being undertaken: leaf-tracking studies would be appropriate if we were interested in rates of damage to a particular species of tree. Since the frass we collect can be derived from any plant above a particular trap, one can usually say little about rates or agents of damage to a particular tree or species.

Funding

The work is funded by the Earthwatch Institute, which as most readers will know, allows a scientist to realize their goals by gathering together adult volunteers, who help with the work and also pay for most of a project's expenses. Earthwatch support is particularly suitable for labor-intensive projects. Frass collecting meets that criterion better than any other method of doing canopy science that I can think of, with the possible exception of litter-fall studies (where frass is often simply ignored).

Contact details

Dr. Eric John Olson, Biology Dept., Wellesley College, Wellesley, MA 02481 USA.
Email: olsonen@rcn.com

Monitoring Biodiversity: Analysis of Amazonian Rainforest Sounds

By
Klaus Riede

THE number of species inhabiting the earth can only be estimated and is a matter of vigorous debate among scientists with estimates ranging from 5 to 80 millions. This is partly due to the fact that so many insect species inhabit tropical forests – especially in the forest canopy (Erwin 1991, Gaston 1991). Due to the accelerated rate of habitat conversion and destruction many of these species will have become extinct before being documented by science and so attempting to inventory the worlds' species is a race against time (Stork 1988). Consequently there is now a need to be able to conduct rapid surveys of forest areas in an attempt to identify the most biodiverse areas (Roberts 1988). Conducting inventories in tropical forests is especially difficult as the complexity of tree architecture provides numerous niches and many species inhabit inaccessible regions of the canopy. Coupled with this, the rarity, excellent camouflage, cryptic lifestyles and nocturnal habits of a great proportion of species makes visual censuses extremely difficult.

The lack of information and the consequent problems associated with biodiversity monitoring has led to the development of 'Rapid Assessment Programmes' (RAP) (Tangley 1992), to identify 'hotspots' of biodiversity and to inventory certain key groups in these critical areas. RAP scientists record the diversity of selected indicator groups which are often by no means sufficient when comparing forests, because different groups react differently to disturbance and other factors (Lawton *et al* 1998). In addition the taxonomic skills required to accurately identify even these indicator groups are labour intensive, costly and increasingly are in short supply.

A considerable number of rainforest creatures indicate their presence acoustically. When walking in a rainforest one can hear a multitude of species even if one cannot see them. Sound analysis through acoustic surveys therefore has the potential to provide an additional window for quick biodiversity assays, particularly in tropical habitats. The example described in this article relates to sampling of the cricket (*Gryllidae*) community of a tropical lowland forest in Ecuador.

Methods

The unaided human ear neither resolves the temporal structure nor the full frequency range of insect songs. Therefore it is necessary to use some form of equipment to aid the human ear. Historically the spectrograph was used and temporal analysis was accomplished by filming oscilloscope tracks (Pierce 1948). However advances in computer technology now allows the use of digital analysis to calculate spectrograms.

Recordings in Ecuador were made with a condenser microphone on a Sony cassette recorder along transects laid out in the forest. These were repeated twice a day at different hours, for two weeks, at each of the sampling locations. The recordings made were then analysed with a spectrum analyser which produces an on-line fast Fourier Transform, visible on a colour monitor.

Results

Frequency bands below 3 kHz are mainly occupied by frogs, birds and mammals, which could be identified by the indigenous people of the area. Short broad-band signals (faint songs with a low repetition rate) were produced by katydids (*Tettigoniidae*), which is probably an anti-predator strategy. Conspicuous, repetitive signals with a narrow-band carrier frequency between 4 and 9 kHz are principally generated by male crickets (*Gryllidae*). Figure 1 gives some examples of graphical displays of songs for typical representatives of major *Orthopteran* groups.

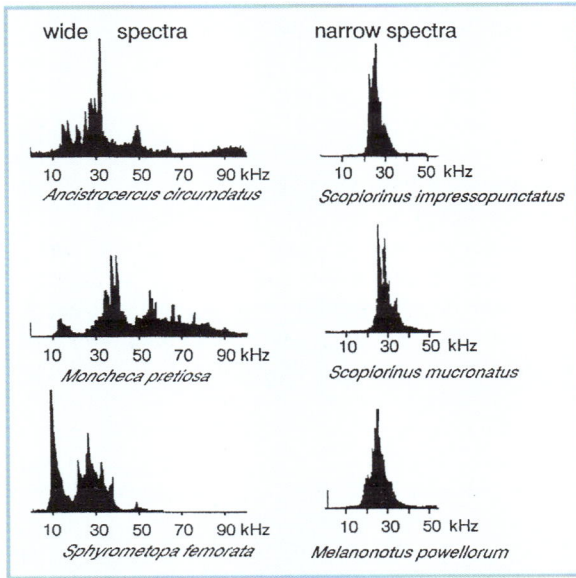

Figure 1 Power spectra of Neotropical *Tettigoniidea* (rearranged after Morris & Beier 1982). Sound amplitude (in relative units on the linear scale) is plotted against frequency. Species on the left have wide frequency spectra, reaching far into ultrasound. Species on the right have narrow carrier frequencies (after Riede 1998, with permission from the author).

This 'acoustical fingerprinting' of different species allows non-invasive mapping to be carried out and considerations about the structure of the 'acoustic community' to be made. To fully describe a community it is necessary to know both the number of species and also their relative abundance. From this, diversity indices can be calculated which combine both species richness and relative abundance into a single statistic. The most commonly used of these is the Shannon-Weiner Index. In Ecuador the index value was H = 2.789, which is rather low when compared with other values obtained when trapping tropical morphospecies, which can reach values as high as H = 5.0 for moths attracted to light in Papua New Guinea (Herbert 1980). One explanation for this could be that reduced diversity has been observed in communities with strong biotic interactions (Caswell 1976). In this case biotic interaction could perhaps be interpreted as competition for acoustic transmission channels.

Discussion

The example discussed here suggests that calling behaviour can be exploited for biodiversity assays in rainforests. The species-specific songs of crickets are an excellent tool for the taxonomist because they evolved for the recognition of conspecifics, and therefore fulfil the very definition of the biological species concept. The acoustic traces recorded at any one location provide a good indication of the number of species. However a comparison of local with regional diversity will only be possible when a larger database of the sounds in different areas exists. Automating the detection and analysis of the sounds could facilitate this. This allows mapping of the different cricket species in the field. A 'cricket detector' is being developed by using neural networks (Dietrich *et al* 2001), which allows more efficient classification and preliminary analysis in the field by displaying relevant parameters like frequency and pulse rate. Songs of other insects like katydids (*Tettigoniidae*) and cicadas (*Cicadidae*) are more complex than cricket songs (for Cicadidae, see Riede 1995). However acoustical analysis is facilitated tremendously by ongoing advances in computer technology, so that this method will also be applicable to these insect groups within the near future.

One of the most attractive elements of this technique is its non-invasive character, compared to fogging, which is widely accepted by the public and conservation agencies. The other is that it can be adapted to monitor hard to access habitats such as tree canopies, where microphones could be pulled up to the top of trees on ropes.

This acoustic inventorying and monitoring of *Orthoptera* species could provide the necessary data for the development of conservation strategies and for monitoring their successful implementation. Once identified, songs are a highly reliable taxonomic feature and allow determination down to the species level. Although this technique has only been applied to crickets, it has potential to be extended to other insects like cicadas (*Cicadidae*). The songs of many others groups are more complex than those of crickets, but with advances in computer technology it will not be too long before this analysis is fast and easy.

Despite these many advantages, the potential of acoustic surveys is still under exploited. One reason for this might be the poor state of insect song documentation. In most cases *Orthopteran* songs are published as spectrograms and/or oscillograms, but unfortunately scales of temporal and spectral resolution differ widely. Ideally songs should be cross-referenced to the recorded specimens which are often undescribed. The Australian Phonotek (CSIRO) fulfils these requirements and has a preliminary labelling system which assigns number codes for species with uncertain taxonomic status (Rentz 1987, Rentz & Balderson 1989). Perhaps the worldwide introduction of this system needs to be considered, especially as there is such a high number of undescribed tropical *Orthoptera* species. A German bioinformatics project is now trying to digitise and pool major sound collections within one 'Virtual Phonothek' (see www.dorsa.de).

Insects are particularly subtle indicators of habitat quality and diversity (Brown 1989); as cricket habitats are strongly determined by local microclimate, these data could serve as a subtle indicator of climate change in the area.

This paper has demonstrated the numerous values of having a virtual Phonotek library of insect songs for biodiversity. Leading on from this the implications of this technique for conservation are also clear.

Contact details

Professor Klaus Riede, Zoologisches Forschungsinstitut und Museum Alexander Koenig, Adenauerallee 160, D-53113 Bonn, Germany. Email: k.riede.zfmk@uni-bonn.de

Literature Cited

Brown H (1989). Conservation of neotropical environments: insects as indicators. *The Conservation of Insects and their habitats*. Collins N &Thomas J. Ed Academic Press, London 349–404

Caswell H (1976). Community structure: a neutral model analysis. *Ecol. Monogr.* 46: 327–354.

Dietrich C, Schwenker F, Palm G (2001). Classification of Time Series Utilizing Temporal and Decision Fusion, In : *Proceedings of Multiple Classifier Systems (MCS) 2001*, Springer, Cambridge 2001, pp 378–387.

Erwin T (1991). How many species are there? Revisited. *Conserv. Biol.* 5: 330–333.

Gaston KJ (1991). The magnitude of global insect species richness. *Conserv. Biol.* 5: 283–296.

Hebert PD (1980). Moth communities in montane Papua New Guinea. *J. Anim. Ecol.* 49: 593–602.

Lawton JH, Bignell DE, Bolton B, Bloemers GF, Eggleton P, Hammond PM, Hodda M, Holt RD, Larsen TB, Mawdsley NA, Stork NE, Srivastava DS & Watt AD (1998). Biodiversity inventories, indicator taxa and effects of habitat modification in tropical forest. *Nature* 391: 72–75.

Morris GK & Beier M (1982). Song structure and description of some Costa Rican katydids (*Orthoptera: Tettigoniidae*). *Trans. Am. Entomol. Soc.* 108: 287–314.

Pierce GW (1948). *The songs of insects*. Cambridge, MA. Harvard University Press.

Rentz DC (1987). Techniques and approaches in studying an unknown fauna: the Tettigoniidae of Australia. In *Evolutionary Biology of Orthopteran Insects* (ed. Baccetti BM), pp. 427–432. Ellis Horwood. Chicester, UK.

Rentz DC & Balderson J (1989). *A Monograph of the Tettigoniidae of Australia. Volume 2, The Austrosaginae, Zaprochilinae and Phasmodinae*. CSIRO, Australia.

Riede K & Kroker A (1995): Bioacoustics and niche differentiation in two cicada species from Bornean lowland forest. *Zoologischer Anzeiger* 234, 43–51.

Riede, K (1998): Acoustic monitoring of Orthoptera and its potential for conservation. *Journal of Insect Conservation* 2, 217–223.

Roberts L (1988). *Science* 241, pp. 1759–1761.

Stork NE (1988). Insect diversity: facts, fiction and speculation. *Biol. J. Linn. Soc.* 35: 321–337.

Tangley L (1992). *Mapping biodiversity: lessons from the field I*. Conservation International, Washington DC.

Vertebrates

By
Hans Winkler

Sampling Birds in the Canopy

MOST ornithological observations are carried out close to the ground, an obvious shortcoming in forests containing tall and dense canopies. Still, the majority of quantitative studies of forest bird communities, particularly in the tropics, are based on understory birds (e.g. Karr 1980, Bierregaard 1990). To get more reliable data on canopy birds, observations from platforms and towers and mist netting in the canopy have been employed. At the Surumoni crane site in Venezuela (see section previously for a fuller description of the site) a combination of methods has been used.

Setting up the nets in the canopy

One of the most effective ways to get quantitative information on the number of birds at a particular location is to set up mist nets. At the Surumoni site more than 100 m of mist nets were used at ground level. Additional net sampling with only a couple of nets was carried out at other sites in roughly a 15 km section along the Orinoco. When we arrived at the crane site we could hardly wait to set up a net in the canopy. We simply cut two sticks jumped into our crane's gondola and rushed into the canopy of some of the tallest trees. There we fastened the tops of each of the sticks onto a branch and strung a pair of thin ropes at the bottom ends of the sticks to fix the nets to more distant hanging points (see Figure 1). In this rudimentary frame a 6 m net was fastened. That first net, number one as it was dubbed, caught a high diversity of birds. In fact, this one and two other similarly established nets caught as many birds as about 50 m of nets in the understory.

Parallel to this, we experimented with nets which were constructed according to the ideas presented by Munn (1991). They were 12 m high and 4 m wide and spread between two heavy wooden horizontal beams. They were custom-made according to our specifications by Vohwinkel, Germany. The first versions were put up with the help of the crane. Later, this net type was used at

Figure 1 Canopy mist net. Photo: courtesy of Hans Winkler.

various places near our crane plot, but outside the crane's reach. To install such a net without the aid of the crane, a thin rope must first be shot over a strong branch of a tall tree, and then the beams with the net are pulled up. The weight of the lower beam produces enough tension to keep the net well spread and to avoid creasing and pockets. It is very important that the ground below the net is cleared and that a heavy-duty plastic tarpaulin is available. The tarpaulin is folded when the net is up and opened out only when the net has to be lowered to the ground for taking out birds or disabling the nets for the night. All the nets were closed during the night to avoid catching bats.

Observing and recording birds in the forest

Observations of canopy birds were done in various ways. First, via the traditional methods of fixed point sampling and slowly walking transects spotting birds with binoculars and telescopes. Second, by using the crane gondola as an observation post. The third method used was climbing the tower of the crane and observing birds in the different forest strata, or sitting on the cranes' boom and watching the birds from there. The advantage of these latter methods is that they can be carried out even when the crane is in use for other research. Since the Surumoni crane was running on tracks, a relatively large area could also be covered. The observations from the boom were particularly useful to spot birds travelling or foraging above the canopy. Observations of some of the more widely ranging frugivores (e.g. parrots, toucans, cotingas) would have been almost impossible without this vantage point. In addition single rope canopy access techniques were used to observe birds at certain fruiting trees outside the crane plot.

Pollination of trees by birds in the canopy is generally poorly studied. A nectivorous parrot from the South Pacific, *Phygiys* from Fiji. Photo: Andrew Mitchell.

All these techniques were used to get quantitative data on the biometry, foraging and habitat use of birds. The opportunity of moving around the forest canopy with the crane was especially important for the study of frugivory. The gondola allowed fruits that were visited by birds to be closely examined, light measurements at many foraging sites to be taken and in some cases valuable data at nest sites to be collected.

The pros and cons of using a crane

The shortcomings of crane-based methods for observing birds are the relatively small area covered by the crane and the corresponding sampling problems (pseudo-replication). However the nets set up with the help of the crane (first method) can be placed in the very top layers of the canopy which is a great advantage. The other net type used covers a larger vertical section of the forest, but can not be put up as high as the former type due to the limitations of using a catapult for inserting a line. The advantage of this net type lies in the fact that they can be employed anywhere that sufficiently tall and strong trees are available without the need for a crane. Platforms and tree climbing can also enable valuable data for specific occasions (such as nest locations, local food sources) to be collected. Although we only had the crane as an observation tower, we think that for certain projects a set of relatively inexpensive observation towers reasonably well spread over a larger area could be very valuable for observing and sampling birds for quantitative studies.

Contact details

Profesor Hans Winkler, Kinrad Lorenz Institut Fur Vergleichende, Verhaltensforschug der Osterreichischen, Akademie der Wissenschaften, Savoyenstrasse 1A, A1160 Wien, Austria Tel: 43 1486 2121. Email: H.Winkler@klivv.oeaw.ac.at

Literature cited

Bierregaard RO Jr. (1990). Species composition and trophic organization of the understory bird community in a central Amazonian terra firme forest. pp. 217–236. In: Gentry AH (ed.), *Four neotropical rainforests*. Yale University Press, New Haven & London

Greenberg R (1981). The abundance and seasonality of forest canopy birds on Barro Colorado Island, Panama. *Biotropica* 13:241–251

Greenlaw JS & Swinebroad J (1967). A method for constructing and erecting aerial-nets in a forest. *Bird-Banding* 38: 114–119

Humphrey PS, Bridge D & Lovejoy TE (1968). A technique for mist-netting in the forest canopy. *Bird-Banding* 39: 43–50

Karr JR (1980). Geographical variation in the avifaunas of tropical forest undergrowth. *The Auk* 97: 283–298

Munn CA (1991). Tropical canopy netting and shooting lines over tall trees. *J. Field Ornithol.* 62: 454–463

Nadkarni NM (1988). Use of a portable platform for observations of tropical forest canopy animals. *Biotropica* 20: 350–351

Nadkarni NM & Matelson TJ (1989). Bird use of epiphyte resources in neotropical trees. *Condor* 91: 891–907

Winkler H & Preleuthner M (2001). Behaviour and ecology of birds in tropical rain forest canopies. *Plant Ecology* 153: 193–202

Sampling Birds in the Forest Canopy using a Canopy Crane

By
David C Shaw

THE major limitation in avian sampling using a crane is that the sample represents only one forest stand (n = 1). Although the crane can be used to replicate within the forest stand, the research question must be limited to single stand phenomena. One solution is to develop access to the canopy in several other forest stands, perhaps using walkways or platforms that allow repeated access at locations with similar potential for observation of the 3 dimensions of the canopy. We have been utilizing and experimenting with a variety of techniques at the Wind River Canopy Crane Research Facility in an old-growth temperate conifer forest. Most our efforts were limited to use of the crane. I describe our experiences here.

The theory of sampling vertical occurrence of birds using fixed area point counts

Fixed area point counts (Hutto *et al* 1986, Manuwal & Carey 1991, Ralph *et al* 1995) are typically used to determine the relative abundance of birds in various habitats. It is usually a ground-based technique that incorporates the entire vertical volume of a vegetation stand in a circular fixed area plot. An observer stands or sits on the ground, in the center of the plot and for a certain time period (between 3 and 20 minutes usually) counts all the birds detected within the circle. A standard circle size used in the forests of the Pacific Northwest of North America is 50 m, although 30 m is also common.

We used fixed area point counts to determine the relative abundance of birds in vertical strata of a 64 m tall, old-growth, temperate coniferous forest (Shaw & Flick 1999, Shaw *et al* in review). The canopy was divided into three strata, low (0–20 m), mid (20–40 m), and upper (40 m to under the jib of the crane, which varies from 67 to 75 m depending on slope position). Four canopy gaps were located in the north, east, south and west directions at the end of the jib (87 meters from the tower). Point count locations were established in the mid zone of each strata, at 10 m, 30 m, and 50 m in each of the four gaps. To sample the birds, we started in the upper canopy, then moved to the mid and lower strata, and repeated the procedure in each of four directions. The point count plot was 30 m in radius, and limited to the canopy strata where the gondola was located. When arriving at a point location, we waited three minutes as a 'quiet down' and the observation period lasted 5 minutes. All birds seen or heard in the plot were noted. A survey day consisted of 12 fixed area point count plots, 4 in the lower canopy, 4 in the mid, and 4 in the upper canopy and took approximately 2 hours. This sampling was done weekly for three years. The data provided us with a relative abundance of birds in each of 3 canopy strata.

The major problem with the technique, in the view of some reviewers, is that the sequence of observations (upper, then mid, then low) does not provide independence of observations. A single bird could be counted twice in the sequence if it descends through the canopy along with the gondola. We have defended the technique because our research question involves, 'Where do bird species occur within the canopy?' and we are interested in knowing if a chickadee is present in the mid and lower canopy. However, the problem can be dealt with in two ways if it is considered insurmountable. One way is to take our data set and randomly choose only one plot location in each descent sequence. This throws out lots of data, but corrects the independence problem. The other way, for those considering this technique, is to only sample one of the locations in a vertical series on any given day. For example, a survey day could consist of only three gap descent locations, and in one, the upper canopy plot could be done, in the other a mid, and in another a lower canopy plot. This also deals with the independence problem.

Depending on canopy stature and access technology, there are many variations that can be done with fixed area point count plots.

Research on birds using the crane
Using the canopy crane for observations on bird foraging:
As a field of research foraging ecology has received significant attention from ornithologists and has a well developed theoretical and technical framework (Morrison *et al* 1990). Observers rarely leave the forest floor however, and the number of foraging ecology studies that utilize canopy access technology has only recently begun to increase (e.g. see Nadkarni 1988, Nadkarni & Matelson 1989, Munn & Loiselle 1995). Given the n = 1 caveats, canopy cranes can significantly improve observations on forest mid and upper canopy bird foraging behaviour, and supplement further reaching studies (Winkler & Preleuthner 2001).

I have been using the canopy crane for an n=1 forest stand study of avian foraging locations at Wind River. This study has emphasized small diurnal songbirds and woodpeckers. The basic idea is to build up an unbiased data set on foraging observations for as many bird species as possible. In an attempt to provide unbiased observations in the upper, mid and lower canopy, I use the crane to hang the gondola for 20 minute intervals at 50 m, 30 m, and 10m, in two vertical sequences, at the end of the 85 m long jib. These sequences are either north-south, or east-west, which means that the vertical sequences are separated by 170 m. At each location I quantify all foraging movements made by birds that I can observe. Data taken on each foraging observation include: date & time, bird species, tree/plant species, tree/plant condition (dead, live), crown class, bird height above the ground when foraging, position of the bird in the tree (bole, branch, branchlet, foliage), crown position, canopy level, foraging technique, substrate, food type and any other notes.

However, the independence of observations must be adhered to in foraging studies, and therefore, the first observation of a given species in any plot within a vertical sequence is the only data point taken for that species. If more than one observation is taken from the three possible within this vertical sequence, the observation used for analysis is randomly chosen. For a given survey day then, only two foraging observations are possible for any given species as the observations have to be separated in space, so adjacent plots can't be used. The survey sequence takes about 2.5 hours. Because the study has been limited to the canopy crane site, I will be submitting the research results to a smaller, natural history focused journal. The n = 1 forest stand focus does limit the utility of foraging ecology studies using cranes, and I would suggest complimenting crane based observations with others locations.

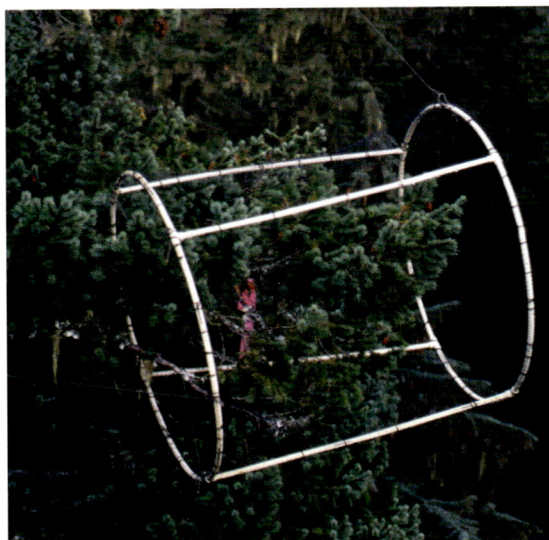

Figure 1 Avian exclosure developed by T. Torgerson. Photo: David Braun.

One problem with this experimental design is that it seems more likely that observations of foraging will be made in the upper rather than in the lower canopy. This is because there is far greater visibility in the upper canopy where the vegetation density is lower. I am not limiting the observations to a specific distance, rather, taking any observation I can. Although I attempt an unbiased approach by vertically stratifying observation locations, I may be biased to upper canopy observations due to this factor.

Using the crane to install avian exclosures.

Canopy cranes are ideal for installing avian exclosures on trees. Avian exclosures are used in a variety of ways to determine the role of birds in controlling insect populations and herbivory (Torgersen, *et al* 1990). They can include whole tree or individual branch exclosures (Campbell *et al* 1981). We have done a preliminary study to determine the effectiveness of avian branch exclosures in large conifers at Wind River. We attempted to use the methods developed by Sunshine Van Bael who has been doing a very rigorous study at the two Panama cranes. We installed fifteen exclosures using wooden dowels and agricultural netting, and found that they did not hold up well when built by us. T. Torgersen is developing an exclosure that is made out of tubular pvc in a circle approximately 1 m in diameter and can be as long as one desires. The unit is built on the ground and placed over a branch by the gondola. It is held in place by a guy wire attached directly above and two wires attached below the unit. When field tested, this pvc frame with netting around it, held up quite well, even through the winter.

Comparison of ground observations vs. gondola observations.

We have used the canopy crane to compare ground based observations of bird occurrence in the canopy with those observed by someone in the gondola (unpublished). This is a unique application of canopy crane technology. The observer stands directly below the gondola (with hard hat on). Using hand signals to determine the start and end time of observation periods, each observer did a fixed area point count plot in the upper, mid and lower canopy. Comparisons can be made of bias by either observer. For example, very quiet bird songs, or non-vocalizing birds, were often missed in the upper canopy by an observer on the ground. Alternatively, the ground observer often detected birds that flew behind the back of the observer in the gondola. Birds that have very loud songs or calls were rarely missed by either observer. The technique seems most useful as a training exercise and a reality check on data collected from ground or canopy.

Final thoughts

The creative ornithologist, with access to a canopy crane, can find them a useful tool. They can be used for stand-alone studies or incorporated into larger studies that require numerous sites. Single forest stand observations and studies from canopy cranes can be used to generate hypotheses that require testing on a larger scale. For example, at the Wind River Crane site, the fixed area point count study has led us to the hypothesis that birds in old-growth forests utilize the upper canopy to a more significant degree during winter, a completely new thought. We are keen to test this hypothesis on a wider scale, and without the crane study, perhaps would have never developed this hypothesis. Although Munn and Loiselle (1995) dismiss canopy cranes as unlikely to improve canopy avian studies, I feel they can play a very useful role.

Contact details

Dr David Shaw, Wind River Canopy Crane Research Facility, University of Washington, 1262 Hemlock Road, Carson, WA 98610, USA. Email: dshaw@u.washington.edu

Literature cited

Campbell RW, Torgersen TR, Forrest SC & Youngs LC (1981). *Bird exclosures for branches and whole trees*. General Technical Report PNW-125. US Department of Agriculture, Forest Service, Pacific Northwest Forest and Range Experiment Station. 10 p

Hutto RL, Pletche SM & Henkdricks P (1986). A fixed-radius point method for non-breeding and breeding season use. *Auk* 103: 593–602

Manuwal DA and Carey AB (1991). *Methods for measuring populations of small, diurnal forest birds*. General Technical Report PNW-GTR-278. Pacific Northwest Research Station, Forest Service, US Department of Agriculture, Portland, Oregon

Morrison ML, Ralph CJ, Verner J & Jehl JR Jr. (ed) (1990). Avian Foraging: Theory, Methodology, and Applications. *Studies in Avian Biology* 13: 1–515

Munn CA & Loiselle BA (1995). Canopy access techniques and their importance for the study of tropical forest canopy birds. Pages 165–177, in: Lowman MD & Nadkarni NM. *Forest Canopies*. Academic Press, San Diego, California, USA

Nadkarni NM (1988). The use of a portable platform to observe bird behavior in tropical tree crowns. *Biotropica* 20: 350–351

Nadkarni NM & Matelson TJ (1989). Bird use of epiphyte resources in neotropical trees. *Condor* 91: 891–907

Ralph CJ, Sauer JR & Droege S (technical ed) (1995). *Monitoring Bird Populations by Point Counts*. General Technical Report PSW-GTR-149. Pacific Southwest Research Station, Forest Service, U.S. Department of Agriculture, Albany, California, USA. 187 p

Shaw DC & Flick C (1999). Are resident songbirds stratified within the canopy of a coniferous old-growth forest? *Selbyana* 20: 324–331

Shaw DC, Freeman EA & Flick C (In review). The vertical occurrence of small birds in an old-growth Doulgas-fir-western hemlock forest stand. *Northwest Science*

Torgersen TR, Mason RR & Campbell RW (1990). Predation by birds and ants on two forest insect pests in the Pacific Northwest. *Studies in Avian Biology* No. 13: 14–19.

Winkler H & Preleuthner M (2001). Behaviour and ecology of birds in tropical rain forest canopies. *Plant Ecology* 153: 193–202

Catching Bats (Chiroptera) in Tropical Forest Canopies

By
Robert Hodgkison,
Ahmad bin Dagu,
Sharon T Balding,
Tigga Kingston,
Zubaid Akbar
and
Thomas H. Kunz

MIST netting at ground level, provides the most commonly used method for sampling bats in tropical forests (Findley & Wilson 1983; Kunz & Kurta 1988; Heideman & Heany 1989). However, recent studies have shown that capture rates and species composition can vary dramatically in relation to net height (Bernard 2001; Cosson 1995; Francis 1994; Hodgkison 2001; Ingle 1993; Kalko & Handley 2001; Zubaid 1994). Hence, without the use of canopy nets, some species of bat may be seriously under-represented in inventories and ecological studies (Francis 1994; Ingle 1994; Kalko & Handley 2001). The aim of this chapter, therefore, is to describe a canopy netting technique, modified from that of Whitaker (1972), which has been used success-fully for the capture of fruit bats (*Megachiroptera: Pteropodidae*) in an old growth lowland dipterocarp forest in Malaysia. Although working on the same principle as its predecessor, this technique offers a number of potential advantages, including:
1. Fewer materials – all cheaply and easily obtained in most parts of the world.
2. Modified "falls" (vertical support ropes) – which prevent pocket loss, by main-taining net tension throughout the rig to heights of a least 30 m.
3. A quick and easy method of net attachment – which allows at least nine nets to be handled simultaneously as a single unit.

Each net rig takes two people up to two days to construct, and costs approxi-mately US$50 (excluding nets and labour). In recognition of the limitations of this technique, further notes are also provided on a canopy netting technique first described by Munn (1991).

Site Selection
To construct the net rig first locate a suitable site. The site must have two sturdy branches, of approximately the same height in the forest canopy, which:
1. Must be separated by a distance that comfortably exceeds the overall length of the mist nets
2. Must have clear space, directly below, in which the nets can be placed without obstruction from vegetation. A small to moderate amount of vegetation clear-ance may be required to achieve the latter, although this should obviously be minimized wherever possible.

Equipment and installation
The basic tools and materials required for the construction of the rig are listed below. The precise specifications, however, may vary according to local availability.

TABLE 1

EQUIPMENT	USE	SPECIFICATIONS
Parang or machete	*to clear vegetation, cut rope, and make posts*	

Fishing line	to guide a thin rope into the canopy	heavy duty e.g. 10lb breaking strain
Fishing reel	to release the fishing line	open-faced spinning reel, with a line capacity of at least 150 m
Fishing rod (base section), or smooth pole	to hold the fishing reel.	
Fishing weight	to add weight to the fishing line	size 8
Catapult or crossbow, plus plenty of rubber tubing	to fire the fishing line up into the canopy	choose a catapult model with a supporting arm brace
Thin rope	to guide a thick rope into the canopy, and to attach 'd' shackles to the anchor posts	approx. 2 mm diameter, preferably braided
Thick rope	to secure the rig in the canopy (i.e. cross-rope), hold the nets in position (i.e. the "falls"), and reduce the net sag in the middle of the stack (i.e. side loops)	approx. 10 mm diameter, preferably braided
Matches or cigarette lighter	to melt the rope ends and prevent fraying	
'D' Shackles, or similar	to guide the "falls"	eight 'D' shackles required per stack
Tarpaulin	to protect the nets, catch splats, and wrap the nets during transit	
Short rubber bungees	to bundle the tarpaulin	
Black nylon thread	to link mist nets along their margins	
Electrical tape	to cover knots and/or faults in the rope that may otherwise snag the net	pale-coloured tape may be preferable, since this can also be used to label and number the nets in the stack
Binoculars and powerful spotlight	to check nets for captured bats	
Safety Equipment: Hard hat, impact-resistant face shield, goggles, thick leather gloves.	to protect head, face, and hands	

When a suitable site has been located, the first stage in rig construction is to fire a weighted fishing line over the branches in the canopy, using a catapult or crossbow. Depending on the height of the trees, this can be done from the ground, or by climbing a nearby tree to gain a better vantage point. A fishing reel, attached to the end section of a fishing rod (which is stuck into the ground or held away from vegetation), must be used to allow the smooth release of the line (Francis 1994; Munn 1991). Once in place, this fishing line can be used to feed two thin guide ropes up into the canopy (Figure 1). These are then used to feed the main cross-rope over the two supporting branches, and, in so doing, raise the falls into position in the gap between. Figures 1 to 3 illustrate the sequence of events. The cross-rope must be tied securely at either end, to two sturdy trees, to hold whole the rig

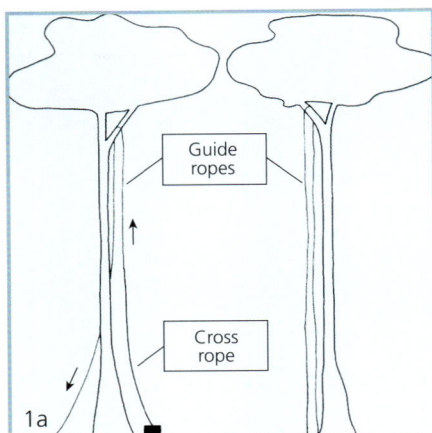

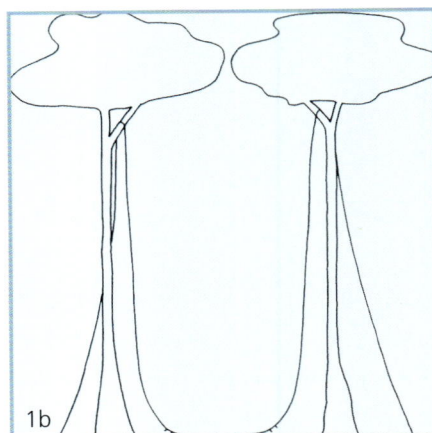

Figure 1a and 1b:
Having gained access to the canopy, using a fishing line and catapult, two thin guide ropes are used to feed both ends of the main cross rope over the two supporting branches.

Figure 2: Paired 'D' shackles are attached to the main cross-rope to carry the falls.

Figure 3: Once threaded through the upper 'D' shackles, the 'falls' are raised into the canopy by pulling on the cross rope at either end.

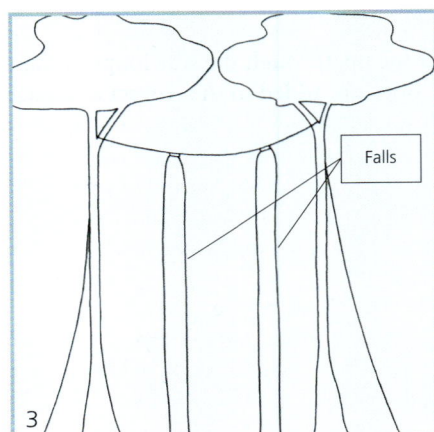

in place. The 'falls' are held in position on the cross-rope by paired 'D' shackles (Figures 2 and 3). These are approximately 30 cm apart and to increase the tension on the net, are attached so that the innermost shackles of each pair are separated by a distance that slightly exceeds the overall length of the mist nets – including the loops. However, this distance should be increased, if there is a large discrepancy between the heights of the two supporting branches (see Hints & Tips, below).

To anchor the falls ropes at the bottom of the rig, four posts must be driven deeply into the ground and 'D' shackles attached with thin nylon chord (Figure 4a and 4b). The falls can then be threaded through the lower 'D' shackles, and secured into position on the inside of the rig, using a combination of two knots – as shown in Figure 5. Since the falls are untied during each net change, it is advisable that the lower knot should be tied relatively loosely at this stage, with plenty of extra slack.

To carry the nets up and down the rig, net carriers can be constructed from thick chord, bound with electrical tape (Figures 6a and 6b). These carriers should sit freely, between the upper and lower knots of the falls, allowing the ropes to rotate freely without twisting. Once the carriers are in place, the nets can then be threaded onto the rig, by unfastening the lower knot of the falls and passing this

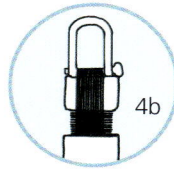

Figure 4a and 4b: After the cross rope is tied securely at both ends, four posts are driven deeply into the ground to carry the lower 'D' shackles.

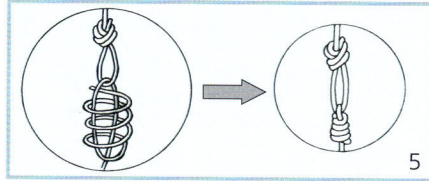

Figure 5: Knots used for securing the 'falls'.
Note – these knots should be tied on the inside rope of the 'falls'.

rope up, through the side loops of the mist nets, taking care to ensure that all the loops are added in the correct sequence! A length of tarpaulin, spread on the ground between the falls, will help to keep the nets free of litter and vegetation, and will also be useful for collecting the faecal splats of fruit bats when the nets are in use. The top loops of the uppermost mist net, which are added last, should be tied directly onto the carriers. The falls ropes can then tightened, taking up the extra slack, and the lower knot fastened securely and wrapped with electrical tape, to reduce the risk of snagging. The nets can then be hoisted into the canopy, by pulling on the falls at both ends of the rig, and joined together along their margins, using short lengths of black nylon thread. To reduce net sag (and, hence, prevent pocket loss), short loops of thick nylon chord can be tied around the falls at every second net (Figure 7a and 7b). All nets should also be labelled and numbered in sequence (starting from the bottom up), to record the capture height of all netted bats.

Once joined, any number of mist nets can be added or removed from a single rig within approximately five minutes by two people. To remove the nets, a short length of rope is threaded through the side loops, at either side of the net – again, taking care to ensure that all loops are included in the correct sequence! The ends of the rope are then tied together securely, to make a continuous loop, and the nets released, by untying the falls. The falls can then be re-fastened loosely, with plenty of slack, and the nets folded and wrapped in the tarpaulin using short rubber bungees. This process is simply reversed, to return the nets to the rig.

Field trials

During a two-year field study in Malaysia (Hodgkison 2001), 24 rigs were constructed within approximately one and a half square kilometres of forest. All rigs were six-metres wide and up to 30 metres tall, with a maximum net area of 162 square metres. Although capture rates varied within each rig, according to net height (Hodgkison 2001; Hodgkison *et al*, in prep.), the average capture rate per 6 x 3 m mist net within each rig ranged from 0.01 to 0.14 bats per mist net hour (0.08 ± 0.04 [mean ± 1 S.D], n = 24). Bat species captured, ranged in body mass from 3.5 to 80.0 g. Other incidental captures included nocturnal birds, flying lizards, frogs, and squirrels, and even a flying lemur! Although some rigs were

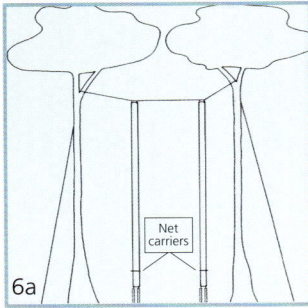

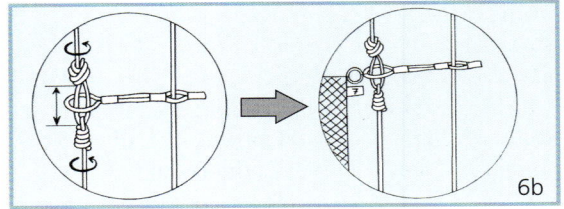

Figure 6a and 6b: The net carriers should sit freely, between the upper and lower knots of the falls, allowing the ropes to rotate freely without twisting.

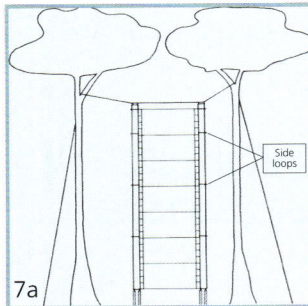

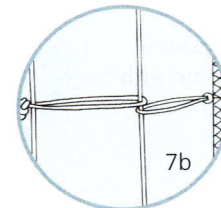

Figures 7a and 7b: To reduce net sag (and, hence, reduce pocket loss), short loops of thick nylon chord can be tied around the falls at every second mist net.

damaged by high winds, which caused the main cross-ropes to snap, most rigs were still fully operational and in good working condition for at least one year.

Limitations

Despite sampling up to 30 m, some species of bat are seldom captured. This is particularly true of those species mainly active above the forest canopy. In the case of fruit and nectar-feeders, above-canopy species can be most effectively captured at fruiting and flowering trees, using a standard technique described by Munn (1991). This technique uses vertically strung mist nets, supported by poles, which are individually hoisted into the canopy on the end of a single pulley rope (Figure 8). 'Canopy mist nets', designed for this technique, are available from Avinet (Rinehart & Kunz 2001), priced $85 (3 x 6 m) and $125 (3 x 12 m), but can also be constructed from standard mist nets, by re-threading the shelf strings (Munn 1991). Above-canopy insectivores, by contrast, can often be captured over rivers, or in open spaces at the forest edge, using ground nets.

As with any other form of mist netting, the canopy net rigs, described in this chapter, are also strongly biased in relation to diet – since most echo-locating insectivorous bat species, of the forest interior, are particularly adept at avoiding mist nets. Unfortunately, no satisfactory method is currently available for catching these species in the forest canopy. However, acoustic monitoring can be used to identify the echolocation calls of certain species (e.g. Kalko 1995; Kalko & Handley 2001; O'Farrell & Miller 1997, 1999).

Conclusion

Although simpler methods of netting (e.g. Munn 1991) can offer similar rates of capture (Francis 1994; Hodgkison 2001; Zubaid 1994), the continuous net coverage of the canopy net rigs, provides a large surface area for the capture of bats and can also reveal details in habitat use which would otherwise remain undetected (Hodgkison 2001; Hodgkison *et al* in prep.). Therefore, despite the extra costs in time and labour, this technique is recommended for all surveys and inventories, and is particularly recommended for any study on fruit and nectar-feeding bats which is concerned with the vertical partitioning of complex forest habitats.

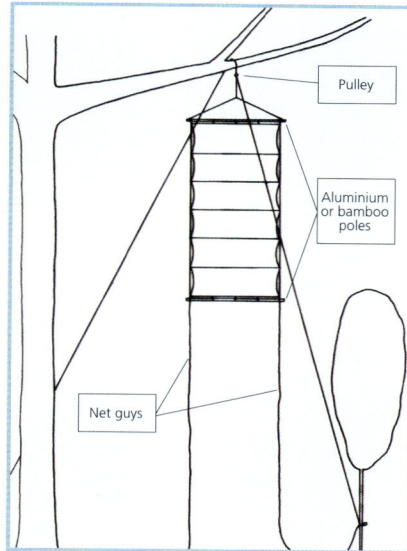

Figure 8: Using a simple pulley, vertically-strung mist nets can be raised into the canopy between two poles. To prevent the net from swaying, particularly when netting for larger species, use wooden stakes to tether the net guy ropes to the ground. This technique is particularly effective for capturing fruit and nectar-feeding bats at their food trees.

Hints & tips

- Practice making low rigs first.
- When firing the catapult or crossbow, slow the release to the fishing line with your hand, as soon as the weight, or bolt, clears the desired branch (Munn 1991).
- Always test the rigs in daylight first. Pay particular attention to the falls. Should twisting occur, you may need to increase the distance between the falls – particularly when there is a large discrepancy between the heights of the two supporting branches. This can be achieved by lowering the rig and re-positioning the upper 'D' shackles.
- Always have binoculars to hand, both during rig construction and when netting at night.
- Use a powerful spotlight at night, and check the nets every 15 minutes for captured bats – lowering the nets if necessary.
- Remove captured bats from the bottom nets first!
- Avoid working on moonlit nights – since bright conditions can drastically reduce capture rates.
- Reduce net shyness (when the bats know that the nets are in place and so avoid them), by rotating between different sampling points.
- Remove the nets from the forest after use – to keep them dry and free of debris, and to avoid opportunistic theft!

Safety

When firing the catapult or crossbow protect head and face at all times with a hard hat and face shield. Thick leather gloves should also be used to protect the hand that holds the catapult (Munn 1991) and to protect against friction burns particularly when pulling on resistant ropes. Remove all jewellery and watches and avoid clothing with excessive zips, buckles, and buttons.

Always beware of falling debris. A hard hat should be worn at all times, and safety goggles are also recommended – particularly when netting at tall fruit-laden

trees. Test the strength of all supporting branches, and be particularly wary of branches heavily laden with fruits and/or epiphytes. Also avoid trees with recently fallen branches!

Never operate canopy nets during high winds, or during and immediately after heavy rain, since this can increase the risk of falling debris. Remove all fishing line from the forest, as this lasts for a long time and can be hazardous to animals. Lastly avoid scratches and bites from captured bats!

Mist net suppliers

AVINET INC., PO Box 1103 Dryden, New York 13053-1103 U.S.A. Tel: Toll free US & Canada (888) 284-6387 or Local/International (607) 8443277. Fax: (607) 844-3915 Website: www.avinet.com
BRITISH TRUST FOR ORNITHOLOGY, BTO, The Nunnery, Thetford, Norfolk, IP24 2PU, UK. Tel: (44) (0) 1842-750050 Fax.: (44) (0) 1842-750030 Website: www.bto.org
ECOTONE, ul. Slowackiego 12, 81-871 Sopot, POLAND. Tel.: (48) 58-552-3373 Fax.: (48) 58-552-1535. Website: www.ecotone.pl

Contact details

Robert Hodgkison* & Sharon Balding, Department of Zoology, University of Aberdeen, Aberdeen, AB24 3TZ, UK
Ahmad bin Dagu, c/o Pos PERHILITAN Kuala Lompat, Kuala Krau, 28050 Mentakab, Pahang, Malaysia
Zubaid Akbar, Department of Zoology, Universiti Kebangsaan Malaysia, 43600 UKM Bangi, Selangor, Malaysia
Thomas Kunz & Tigga Kingston, Department of Biology, Boston University, Boston, Massachusetts 02215, USA
*Corresponding author: Email: rhodgkison@hotmail.com

Literature cited

Bernard E (2001). Vertical stratification of bat communities in primary forests of Central Amazon, Brazil. *Journal of Tropical Ecology* 17: 115–126

Cosson J (1995). Captures of *Myonycteris torquata* (Chiroptera: Pteropodidae) in forest canopy in South Cameroon. *Biotropica* 27 (3): 395–396

Findley JS & Wilson DE (1983). Are bats rare in tropical Africa? *Biotropica* 15: 299–303

Francis CM (1990). Trophic structure of bat communities in the understorey of lowland dipterocarp rain forest in Malaysia. *Journal of Tropical Ecology* 6: 421–431

Francis CM (1994). Vertical stratification of fruit bats (Pteropodidae) in a lowland dipterocarp rain forest in Malaysia. *Journal of Tropical Ecology* 10: 523–530

Heideman PD & Heaney LR (1989). Population biology and estimates of abundance of fruit bats (Pteropodidae) in Philippine sub-montane rain forest. *Journal of Zoology (London)* 218: 565–586

Hodgkison R (2001). *The ecology of fruit bats (Chiroptera: Pteropodidae) in a Malaysian lowland dipterocarp forest, with particular reference to the spotted-winged fruit bat* (*Balionycteris maculata*, Thomas). Unpublished Ph.D. dissertation, University of Aberdeen.

Ingle NR (1993). Vertical stratification of bats in a Philippine rainforest. *Asia Life Sciences* 2: 215–222

Kalko EVK (1995). Echolocation signal design, foraging habitats, and guild structure in six neotropical sheath-tailed bats (Emballonuridae). *Symposia of the Zoological Society of London* 67: 259–273

Kalko EVK & Handley CO Jr (2001). Neotropical bats in the canopy: diversity, community structure, and implications for conservation. *Plant Ecology* 153: 319–333

Kunz TH & Kurta A (1988). Capture methods and holding devices. Pp. 1–29 in Kunz TH (ed.). *Ecological and behavioural methods for the study of bats.* Smithsonian Institution Press, Washington, DC

Munn CA (1991). Tropical canopy netting and shooting lines over tall trees. *Journal of Field Ornithology* 62: 454–463

O'Farrell MJ & Miller BW (1997). A new examination of echolocation calls of some neotropical bats (Emballonuridae and Mormoopidae). *Journal of Mammalogy* 78: 954–963

O'Farrell MJ & Miller BW (1999). Use of vocal signatures for the inventory of free-flying neotropical bats. *Biotropica* 31: 507–516

Rinehart J & Kunz TH (2001). Preparation and deployment of canopy mist nets made by Avinet. *Bat Research News* 42:85–88

Whitaker AH (1972). An improved mist net rig for use in forests. *Bird-banding* 43:1–8

Zubaid A (1994). Vertical stratification of pteropodid bats in a Malaysian lowland rainforest. *Mammalia* 58: 309–311

Acknowledgements

We would like to express our thanks to the Lubee Foundation Inc., Florida, for sponsoring our work on Malaysian fruit bats. We would also like to thank the Department of Wildlife and National Parks, and the Economic Planning Unit of the Prime Minister's Department, for granting us permission to work in Malaysia.

8 Remote Sensing

Large-scale Aerial Photographs

By
Valérie Trichon
and
Daniel
Guillemyn

THE method in itself (i.e. aerial photography) is not new. Large-scale aerial photographs (i.e. with a scale above 1:5000) have been used in tropical forests for about 30 years. However their use is rare, and mainly for the purposes of recognition of several commercial tree species. Several scientists at the Laboratoire d'Ecologie Terrestre (Toulouse, France) have acquired ten years experience studying forest dynamics, structure and diversity in the canopy with the aid of aerial photos. One member in our team has actually been practising aerial photography over forests in France for the last 25 years. This remote sensing method has always been used in conjunction with other field methods for the assessment of forest diversity and structure (tree measurements and botanical samples). Some of the study sites were permanent plots with an available database on tree inventories (Paracou, French Guiana).

Equipment and installation

An aircraft has to be used for getting above the canopy with the photographic equipment. This can be a hot airship (Figure 1) helicopter or Ultra light aircraft. Silver balloons made of mylar filled with helium is another method which we have found to be very useful for marking reference points in the canopy. These are tethered to the ground with a nylon line and then emerge through the trees above the canopy. They are useful as markers both during the flight and afterwards to aid assembly of the mosaic of photographs.

A platform is attached to the aircraft and the various items of equipment are then in turn attached to it. The cameras which we commonly use are 24 x 36 cameras (FM2 Nikon® with a 35 mm

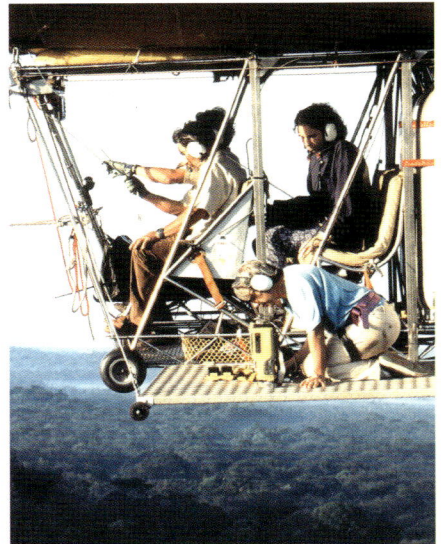

Figure 1 Researchers taking aerial photos whilst suspended beneath the dirigible. Photo: L. Pyot.

lens) filled with black and white negatives or colour slides. As well as this we use a digital camera (Kodak DCS 420, with a 20 mm lens).

Holes were made in the platform (aluminium or wood) to contain the cameras' lens, then the platform was fixed to the aircraft (balloon or helicopter). This process took a couple of days of happy pottering in warmth and moisture to complete.

The decision as to which methods to use was influenced by a number of factors:

1. Aircraft

   We had the opportunity to test the hot-air airship in Cameroon and French Guiana. Afterwards, we wanted to test the ability of other aircraft, such as helicopters and lighter than air craft (LTAC), for flying at low altitude above the canopy. The LTAC was cheaper than the helicopter but the helicopter was more stable and secure. However, use of a helicopter may cause some harmful effect on natural communities due to the down draught caused by the rotor blades.

2. Photographs

   Prior to our acquisition of a digital camera, black and white photographs were preferred to colour slides in remote study sites as the photographs could be developed quickly and easily in the field. A digital camera is very useful as one can immediately check the success of any photographs taken. Whenever possible we use two cameras at the same time, as a precautionary measure (in case there is a problem with one of them).

Useful points

If possible choose to fly at 12 o'clock to minimise the shadow effects on the photographs, but be aware of turbulence, mist and clouds occurring in the middle of the day in tropical countries. Electronic cameras tend to breakdown with vibration, heat and moisture in tropical regions. So always keep a place for a good old reliable mechanical camera. Preparation of flight, security and cameras, must be conducted with great care and precision. If you are new to this technique then use checklists. When everything is ready, be sure that the lens caps of the cameras are removed.

The tiniest details can cause a photographic mission to fail. Thus you should not rely blindly on high technology, such as G.P.S. (Global Positioning System). A conventional altimeter is often the most fail safe method of being sure of your altitude. With balloons used as landmarks, it is essential that their control from the ground is carefully monitored. Once the position and number of balloons has been fixed, make sure that nobody modifies it, for example by adding balloons without having informed the crew. Discovering twice the expected number of balloons glittering everywhere above the canopy is really disturbing.

Research projects

The main research projects we have carried out using this technique concentrated on tree species identification, forest structure and dynamics. In addition to these broad themes aerial photography is good for studies on crown architectural development in the canopy and crown phenology.

We have successfully used aerial photos in a variety of locations, including Cameroon (1991), Sumatra (1994), Guiana (1995, 1996, 1997), as well as in continental France (1973–1995).

At the present time we have not developed any standard protocols for research conducted with this technique. However we do feel that there is a need to create these in the future in collaboration with other researchers in other countries.

Costs

The main costs are associated with staff time, as a well trained and dedicated pilot and a photographer are required at all times. One also has to rent an aircraft, which in French Guiana costs about 960 Euro per hour for a helicopter. Other resources required in the field are cameras, films and a computer (to enable quick viewing of digital images).

Maintenance costs are considerably reduced in our case as the aircraft are rented. This also removes some of the liability from us as this is covered by the rental company. The only major adaptation needed is the secure installation of platforms for the equipment to the sides of the aircraft.

Funding

A regional program in the Toulouse region of France financed the acquisition of a digital black and white camera and a study of the mangrove forest in French Guiana. The main photographic missions were undertaken during the 'Opération Canopée' missions and benefited from associated funding provided by Elf in Cameroon (1991) and Pro-Natura international in French Guiana (1996).

Safety

The main safety consideration is that all procedures are double checked before any flight. It is especially important that the photographer is securely attached to the aircraft itself as he will be sitting on the platform which is bolted to its' outside. Also during any flight he will be concentrating intently on the cameras and the ground control points and thus may not be that observant as to his surroundings. In terms of atmospheric conditions it is not safe to use the aircraft in rain, wind or mist and at the same time these conditions are very poor for photographic image quality.

Contact details

Valérie Trichon & Daniel Guillemyn
Laboratoire d'Ecologie Terrestre, B.P. 4072, 31029 Toulouse cedex 4, France
Email: trichon@cict.fr

Characterising Forest Structure by Airborne Laser Scanning

By
Ross Hill
and
David Gaveau

TRADITIONAL field-based forest inventory and measurement techniques can be time-consuming and expensive if applied at a landscape level. Remote sensing can generate synoptic data rapidly, allowing landscape-scale ecosystem patterns to be discerned. Airborne Laser Scanning (ALS) is an 'active' remote sensing technique that supplies geo-referenced digital elevation data at a high sampling density and with high accuracy. ALS is currently unique amongst remote sensing techniques in that it can record directly the 3-dimensional structure of a forested landscape at a high spatial resolution. This is a non-intrusive method of acquiring information on forest canopy characteristics, and so can be used in areas where access is limited, prohibited or hazardous.

ALS has many advantages when compared with other forest measurement techniques or alternative remote sensing technologies. For example, compared with forest mensuration techniques, ALS offers the ability to record forest canopy information for large areas, quickly, remotely and at high accuracy. ALS is less restricted than aerial photography by weather conditions and sun angle considerations (being operational by day or night), and has the potential to penetrate forest canopies to give information on underlying terrain. Compared with microwave remote sensing (i.e. RADAR), ALS is a more direct and typically more accurate means of deriving canopy height information. However, ALS is currently more expensive and less readily available than aerial photography, and there is at present no space-borne equivalent (such as there is for RADAR) for regional-scale forest inventory.

The technology

ALS technology relies on a laser range finder and a scanning mechanism to measure the elevation at points within a swath beneath the flight-path of an aircraft. Control and recording units on-board the aircraft (which can be either a small aeroplane or a helicopter) regulate the process of on-line data acquisition. To identify the 3-dimensional position of each ranged point, the aircraft must have an integrated position and orientation system consisting of a differential global positioning system (dGPS) and an inertial measurement unit (IMU). In addition, a field-based GPS station has to be deployed in the area of survey at the time of ALS data acquisition.

What does an ALS system record?

ALS operates on a principle called Light Detection And Ranging (LiDAR). A pulse of near infrared laser light is fired at the ground by an aircraft-borne laser scanner. The timing and intensity of the return pulse (following reflection from a feature on the Earth surface) are recorded and used to derive a ranging measurement. At the same time, the IMU component of the ALS system records the roll, pitch and heading of the aircraft to determine its orientation, whilst the dGPS records its precise location. During post-flight data processing, the position and orientation data are used to geo-reference the ranging measurements. Each incident laser pulse

Figure 1 shows a schematic example of a returned waveform from an incident laser pulse over a forested area.

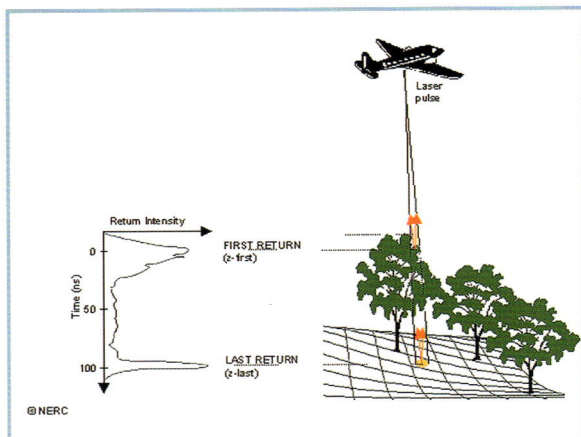

Figure 2 Schematic of laser footprint distribution on the ground for a profiling ALS system (a), and a mapping ALS system (b).

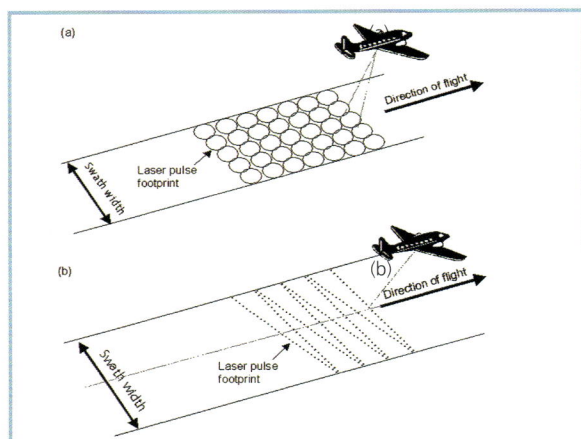

thus supplies accurate (in x-, y- and z-) point-sample elevation data for the ground surface and/or objects on it.

This shows a multiple echo, as the laser pulse will be able to penetrate through at least part of the forest canopy. The width of the returned waveform will depend on the penetration depth of the incident pulse, which will be a function of incidence angle and canopy density. The intensity of the received return pulse from a given depth in the canopy will depend on the amount of near infrared laser illumination penetrating to that depth (by direct penetration, transmission through foliage, or forward-scattering from foliage) and on the reflectivity and orientation of the intercepting surfaces at the wavelength of the laser. In the schematic example shown in Figure 1, the first significant 'echo' in the return pulse (z-first) records information from the canopy surface, whilst the last significant 'echo' (z-last) records ground information. In reality, both the first and last significant 'echo' in the return pulse are likely to record information from within the canopy rather than directly from the canopy surface or forest floor.

How do ALS systems differ?

There are two distinct ways in which the ALS systems used in terrestrial applications record the information of a return pulse waveform:

- *Profiling* scanners such as SLICER (Scanning Lidar Imager of Canopies by Echo Recovery) and LVIS (Laser Vegetation Imaging Sensor) incorporate a waveform digitiser. This records the entire return waveform providing a finely resolved measure of the vertical distribution of plant matter throughout the canopy. With a vertical resolution of up to 0.11 m, this can enable closely spaced canopy layers and potentially the underlying ground to be distinguished within each laser pulse footprint. Scanning a narrow swath of up to 75 m directly beneath the aircraft, the circular laser pulse footprints are 5–15 m in diameter (Figure 2a).

Adjacent footprints are typically contiguous or overlapping, providing a measure of average canopy structure. Where the laser penetrates to the forest floor, a measure of canopy height can be obtained and the relative strength of the canopy and ground returns provides information on 'canopy closure'. Large-footprint laser scanners have thus been used to

Figure 3 The z-first DSM (a) and z-last DSM (b) of Monks Wood, Cambridgeshire, UK, with an example section highlighted. Note the lower homogeneity in canopy elevation in the z-last DSM (b1) compared with the z-first DSM (a1).

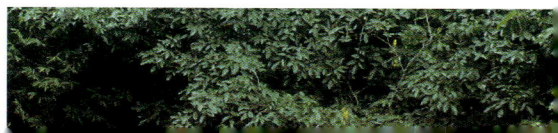

derive forest canopy height profiles (e.g. Harding *et al* 2001), and forest stand characteristics such as height, basal area and biomass (Means *et al* 1999), leaf area index (Lefsky *et al* 1999) and tree volume (Nelson *et al* 1988).

• *Mapping* scanners such as ALTM (Airborne Laser Terrain Mapper) and TopoSys (Topographische Systemdaten), record only the first and/or last significant part of the return pulse (determined by an intensity threshold). The footprint size of each laser pulse on the ground is approximately 0.2–0.3 m diameter. Scanning either side of the central flight line can supply point-samples over a 250–750 m swath, depending on the scan angle and operating altitude. By scanning in sweeps perpendicular to the flight-line, the forward motion of the aircraft generates a saw-tooth pattern of point samples (Figure 2b). With high pulse-repetition laser scanners, thousands of ranging points can be recorded per second resulting in sampling densities of up to 10 laser hits per m^2. This sampling density is adequate to render a distinguishable 3-dimensional model of individual tree crowns. If both the z-first and z-last elevation data are recorded (supplying information from near the canopy surface and forest floor), by a process of interpolating elevation between the point-samples, morphological filtering and data subtraction it is possible to generate a Digital Terrain Model (DTM) of the forest floor and a Canopy Height Model (CHM) of the forest. Small-footprint laser scanners have thus been used to map tree height (e.g. Næsset, 1997a, Magnussen *et al* 1999), and timber volume (Næsset, 1997b).

Large-footprint scanners such as SLICER and LVIS thus provide a better profile of forest vertical stratification, but their practical applications for landscape-scale mapping are limited by poor spatial coverage compared with small-footprint scanners such as ALTM or TopoSys. In forestry applications, large-footprint laser scanners are likely to be used to provide transects of canopy profile data, whilst small-footprint laser scanners are likely to be used for blanket-coverage forest canopy mapping.

A case study: modeling canopy height for a temperate deciduous woodland using a mapping ALS system

Monks Wood (52°24' N, 0°14' W) is a 162 ha area of deciduous woodland in Cambridgeshire, UK. It was established as a National Nature Reserve in 1953. The

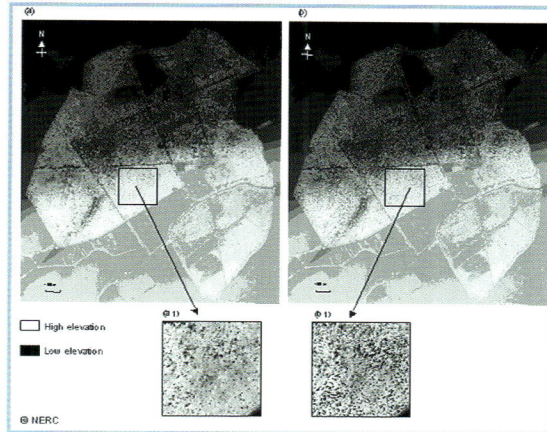

Figure 4
The 1 m spatial resolution Digital Terrain Model (DTM) of Monks Wood, Cambridgeshire, UK.

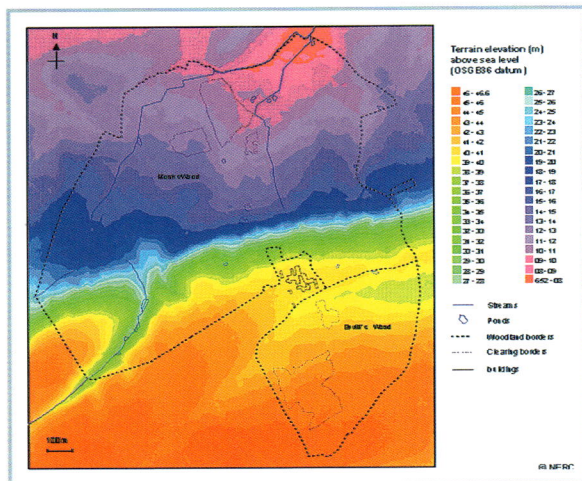

Figure 5
The 1 m spatial resolution Canopy Height Model (CHM) of Monks Wood, Cambridgeshire, UK.

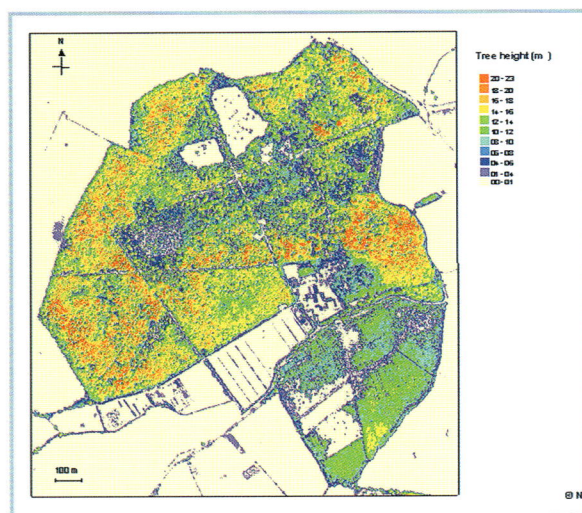

elevation of Monks Wood ranges from 6 m to 46 m, with a maximum slope angle of 14.5°. The main upper-storey tree species are ash (*Fraxinus excelsior*), oak (*Quercus robur*), field maple (*Acer campestre*), aspen (*Populus tremula*) and a few remaining elm (*Ulmus carpinifolia*). Ash is the most common and widespread tree species in Monks Wood. Oak occurs less frequently because of intense felling during the First World War. Field maple is found scattered throughout the woods, whilst aspen and elm form occasional clusters on the wetter soils. The dominant shrub species making up the under-storey and woodland fringes are hawthorn (*Crataegus monogyna*), hazel (*Corylus avellana*), blackthorn (*Prunus spinosa*), dogwood (*Cornus sanguinea*) and wild privet (*Ligustrum vulgare*).

The ALS data

An Airborne Laser Terrain Mapper (Optech ALTM 1210) was flown over Monks Wood on 10 June 2000. Laser pulses were emitted by the ALTM with a wavelength of 1047 nm. A scan angle range of was selected. The parallel flight-lines had over-lapping swaths, resulting in an irregular distribution of points. On average, 1 point was recorded every 4.83 m^2 across the study site. However, the average point-sample density varied between 1 point per 6.50 m^2 in areas recorded in one swath and 1 point per 2.80 m^2 in areas sampled in overlapping swaths. Both z-first and z-last elevation data were recorded for each laser pulse, which generated a circular footprint on the Earth surface with a diameter of *ca* 0.25 m directly underneath the aircraft. The x- and y- position of each scanned point were supplied in British National Grid (BNG) co-ordinates, whilst the z- position was supplied as elevation in metres above an Ordinance Survey 'sea level' datum (OSGB36). The x-, y- and z- data all had a gradation in millimetres.

Pre-processing the ALS data

The ALS data from the ALTM 1210 were supplied as *ascii* tables of x-, y-, z- (and intensity, i) point-sample data for both z-first and z-last return. The elevation information in both the z-first and z-last data was composed of *geomorphologic* height, describing the terrain and *morphologic* height, which related to the height of the vegetation canopy. The focus of the ALS data pre-processing was thus to render an interpolated surface (Digital Surface Model – DSM) from both the z-first and z-last point-sample elevation data, and to separate from these a Digital Terrain Model (DTM) of the forest floor and a Canopy Height Model (CHM) of the forest.

Generation of digital surface models

The creation of DSMs for both the z-first and z-last elevation data involved rendering an interpolated surface from the ALS point-sample data. This first required creating a GIS layer of points in 2-dimensional space (each with an elevation attribute). A Triangulated Irregular Network (TIN) was constructed around the points (Figure 4). A rectangular grid of pixels was then extracted from the TIN with a chosen constant sampling interval in the x- and y-direction of 1 m. This process was repeated twice, creating a 1 m spatial resolution DSM for both the z-first and z-last elevation data (Figure 3).

Generation of a digital terrain model

A process of morphological filtering was applied in which local elevation minima in the z-last DSM were identified and compared with their neighbours to eliminate pixels unlikely to represent ground information. A terrain surface was generated by interpolation from the selected local elevation minima (Figure 4). This processing chain was operated as a two-pass process to reduce the likelihood of interpolating a terrain surface from pixels that did not represent ground information.

Generation of a canopy height model

The CHM was created by per-pixel subtraction of the DTM from the z-first DSM. Thus ± elevation errors in the DTM were transferred to the height values of the CHM (Figure 5).

Important data considerations

The ALTM scanner used for data acquisition in this study was supplied by Optech Inc. of Canada, and operated by the Environment Agency (EA) of England and Wales. The data were supplied by the EA in *ascii* format, and the pre-processing to generate a DTM and CHM was carried out at CEH Monks Wood by the authors (using in-house software and ArcInfo GIS package).

The operating accuracy of the ALTM 1210 scanner is known to be ± 0.15 m in the vertical plane. Validation of the DTM and CHM was carried out by extensive ground survey using an electronic total station (i.e. theodolite). The DTM was found to predict terrain height on average to within 0.16 m (standard error = 0.61 m), whilst the CHM was found to underestimate shrub canopy height by approximately 1 m and tree canopy height by approximately 2 m. The possible causes of these errors in elevation and canopy height include:
- the operating precision of ± 0.15 m for the Optech ALTM 1210 scanner,
- variation in laser pulse penetration into the canopy before reflecting a first significant return (i.e. z-first signal),

- an insufficient ALS point-sampling density to hit all the peaks and troughs in the upper canopy, resulting in a generalised interpolated surface (z-first DSM),
- errors introduced by the processes of extracting a DTM and CHM from the interpolated z-first and z-last DSMs.

Where to find out more information about Airborne laser scanning

The best on-line source of information on Airborne Laser Scanning is the web site at www.airbornelasermapping.com which is run by Martin Flood. This provides a good overview of the research and industrial applications of ALS, and has regularly updated news and demonstration material. The site currently has an index (with internet links) for 68 organisations classified into service providers, system manufactures, consultants, and research/government institutions. There are a number of service providers worldwide, meeting either dedicated surveying needs or consultancy on an individual project basis. The market for ALS data is most developed in North America, and there is now considerable competition in USA and Canada from service providers. There are a similar number of service providers across western Europe as in the USA (notably in the UK and Germany), and a more limited number in Brazil, South Africa, Australia, Japan and Russia. The costs of data provision will vary between organizations and, of course, will be dependent on the area of surveying required. Service providers will also differ in terms of the post-processing resources available to derive value-added products, such as automated DTM generation, feature extraction and classification, and integration with other airborne remote sensing data-sets (such as multi-spectral imagery).

Contact details

Dr Ross Hill, Section for Earth Observation, Centre for Ecology and Hydrology, Monks Wood, Huntingdon, Cambridgeshire, PE28 2LS, UK
Note: Airborne Laser Scanning ©NERC.

Literature cited

Harding DJ, Lefsky MA, Parker GG & Blair JB (2001). Laser altimeter canopy height profiles: methods and validation for closed-canopy broadleaf forests. *Remote Sensing of Environment,* 76, 283–297

Lefsky MA, Cohen WB, Acker SA, Parker GC, Spies TA & Harding DJ (1999). Lidar remote sensing of the canopy structure and biophysical properties of Douglas-fir Western Hemlock forests. *Remote Sensing of Environment,* 70, 339–361

Magnussen S, Eggermont P & LaRiccia VN (1999). Recovering tree heights from Airborne Laser Scanner data. *Forest Science,* 45, 407–422

Means JE, Acker SA, Harding DJ, Blair JB, Lefsky MA, Cohen WB, Harmon ME & McKee WA (1999). Use of large-footprint scanning airborne lidar to estimate forest stand characteristics in the Western Cascades of Oregon. *Remote Sensing of Environment,* 67, 298–308

Naesset E (1997a). Determination of mean tree height of forest stands using airborne laser scanner data. *ISPRS journal of Photogrammetry and Remote Sensing,* 52, 49–56

Naesset E (1997b). Estimating timber volume of forest stands using airborne laser scanner data. *Remote Sensing of Environment,* 61, 246–253

Nelson R, Krabill W & Tonelli J (1988). Estimating forest biomass and volume using airborne laser data. *Remote Sensing of Environment,* 61, 246–253

The Contribution of Remote Sensing to Forest Canopy Science

By
Clare S Rowland
and
Terence P Dawson

FORESTS are of great importance due to their role in the global cycling of carbon, water and energy. From the local to the regional scale they are economically, ecologically and culturally important as sources of raw materials, habitats, recreational areas and for watershed protection (Miller 1992). Consequently, it is important to monitor forest resources at a range of scales from local through to global. This scale and density of monitoring is very difficult to achieve over large areas using ground based measurement techniques, which tend to produce point measures of properties, rather than areal estimates. Point based measurements may be averaged or extrapolated to produce areal values, however the process of 'up-scaling' may not be consistent (Sellers *et al* 1996). Remote sensing has the ability to provide raw data on forest properties at a global scale (Peterson *et al* 1987; Graetz 1990). The problem is to collect, process and derive appropriate information about vegetation with an accuracy, over different spatial and temporal resolutions, which is of use to those in potential user communities, from local forest management (Gopal & Woodcock 1996) to ecosystem and climate modelling (Hall *et al* 1995).

Optical remote sensing of vegetation

Remote sensing is the process of acquiring data, often in the form of images, from a distance. The most well known form of this is satellite images in weather forecasting. Remote sensing can also be used to monitor other environmental phenomenon, processes and environments, including forests. The key advantage over other monitoring techniques is that large areas can be monitored rapidly and non-destructively.

Optical remote sensing instruments measure reflected radiation, much as our eyes do, although sensors can measure wavelengths beyond the range visible to humans. The signals observed, from forest canopies by sensors at visible wavelengths, are influenced by the amount of chlorophyll in the forest canopy, which is in turn related to the ability of the plants to photosynthesize. Vegetation canopies have relatively low reflectance at visible wavelengths due to the strong absorption of light by photosynthesis. In the near-infrared part of the electromagnetic spectrum reflectance is controlled primarily by cell structure, whilst at middle-infrared wavelengths water is a key influence on canopy reflectance. The near-infrared is characterised by high vegetation reflectance in a region called the near-infrared plateau, which is generally flat. Additionally, the near-infrared plateau contains a number of narrow absorption features, due to leaf biochemicals such as lignin and cellulose (Curran *et al* 1997), but the narrowness of the features precludes them from easy observation.

Remote sensing can potentially provide useful information for forest canopy science in a wide range of ways, including:
• Monitoring the spatial extent of forest cover;
• Estimating biophysical variables, such as Leaf Area Index;

- Parameterising or validating ecosystem process models;
- Remote sensing of foliar biochemistry;
- Monitoring and/or predicting forest fires;
- Vegetation stress detection ;
- Estimating forest biodiversity ;
- Monitoring temporal change in forests.

Monitoring forest spatial extent

The key to defining forest spatial extent is being able to discriminate between forest and non-forest accurately. This may require the use of a range of information, besides the basic spectral information provided by the sensor, with temporal information, multi-view angle (Sandmeier & Deering 1999), textural/spatial (Treitz & Howarth 2000) or multiple sensor data (Mayaux *et al* 2000) all being possibilities. This is particularly true if a range of forest types is required, such as lowland rainforest, swamp forest, secondary forest, forest-savanna mosaic and plantation forest.

Estimating forest biophysical variables

Vegetation biophysical properties are important because of their impact on the processes of radiation absorption, rate of evapotranspiration and precipitation interception and loss (Sellers *et al* 1997). Traditional remote sensing practice has tended to use empirical techniques, particularly vegetation indices, to estimate forest biophysical canopy variables. A vegetation index is a simple combination of surface reflectance observed at different wavelengths. The most commonly used is the Normalised Difference Vegetation Index (NDVI). It is designed to exploit the very high reflectance of vegetation at near infrared (NIR) values and the very low reflectance of vegetation at red wavelengths. Field observations are then required to relate the vegetation index value to some variable of interest, such as leaf area index (LAI), canopy cover or the fraction of photosynthetically active radiation (fPAR), via a regression relationship.

An alternative method is to use mathematical models, describing the interaction of radiation with the canopy optical and structural properties, to estimate LAI. Models may be run in a forward mode, where a model is parameterised with values describing the canopy characteristics and the model is run to calculate canopy reflectance. To extract information from remote sensing data, a more useful way of using the model is in inverse mode. In this case the model receives, as input, the widest possible range of canopy reflectance observations, usually from a number of different wavelengths and viewing directions. The model is then inverted using an inversion technique, typically a numerical optimisation algorithm, neural network or look-up table, to provide estimates of the canopy parameter values most likely to produce the observed canopy reflectance.

Vegetation remote sensing is entering a new phase, with a range of quantitative vegetation products, including LAI and fPAR, becoming routinely available at global scale, via the MODerate resolution Imaging Spectrometer (MODIS) and the Multi-angle Imaging SpectroRadiometer (MISR) programs.

Ecosystem process models

Recently, ecologists have been developing ecosystem models to perform assessments of carbon balance and nutrient cycling of forest ecosystems at scales from

regional to global (Running & Coughlan 1988; Cropper & Gholz 1993). Based upon scale-independent deterministic processes and usually relying on meteorological as well as canopy bio-physical/chemical variables, such as LAI and foliar biochemical concentrations, these models can provide large-area assessments of photosynthesis, carbon storage, litter decomposition and evapotranspiration (Woodward 1987). A number of models include detailed canopy physiology, energy exchange, and microbial processes to examine seasonal CO_2 exchange and geographical patterns of annual net primary production (Running & Coughlan 1988; Bonan 1991). For example, FOREST-BGC (Running & Coughlan 1988) estimates annual foliage photosynthesis and foliage, stem and root respiration, taking into account solar irradiance, temperature, humidity, CO_2, nitrogen, and soil moisture limits to photosynthesis and temperature effects on respiration at a daily temporal resolution. Plummer (2000) identifies four methods in which remote sensing data can inform the use of ecosystem process models, namely:
1. To estimate canopy variables (e.g. LAI) to characterise or drive the models.
2. To test, validate or verify model predictions.
3. To update or adjust ecological process model predictions.
4. As a tool to aid understanding of remote sensing data.
Remote sensing efforts to date have tended to focus primarily on method 1 and to a lesser extent on method 2.

Remote sensing of foliar biochemistry

The foliar biochemical concentrations of a forest canopy can be used as an indicator of critical ecosystem processes, for driving and validating ecosystem models (Curran 1994). For example, lignin and cellulose, which comprise the cell walls of leaf tissue, control leaf litter decomposition and hence, nutrient cycling and availability of foliar nitrogen and chlorophyll concentrations determine the maximum photosynthetic rate and thereby the rate of carbon assimilation and net primary productivity (Aber *et al* 1993). Fine spectral resolution remotely sensed data can provide estimates of the concentration of foliar biochemicals in forest canopies. A number of studies, using airborne spectrometers, have successfully estimated the foliar concentrations and content of chlorophyll, nitrogen, cellulose and lignin using empirical techniques (Aber 1994; Curran & Kupiec, 1995), although further research issues remain outstanding (Dawson *et al* 1999).

Forest fires

Remote sensing of fires can be undertaken directly, with the heat of the fires being observed at thermal wavelengths, or indirectly through observation of the smoke at visible wavelengths. Thermal observations (heat measurements) are more appropriate for determining the size, location and intensity of fires, whilst optical data is more suitable for tracking pollution pathways of smoke and particulate emissions. Being able to accurately quantify the intensity of combustion also enables accurate estimates of emission products (e.g. greenhouse gasses, smoke and aerosol particles). Data from the MODIS sensor is now being processed to produce frequent global fire (thermal anomaly) products at a 1km grid scale. The MODIS thermal anomaly products can be viewed at http://edcdaac.usgs.gov/modis/dataprod.html.

Alternatively, remote sensing data can be used to assess the susceptibility of areas to fire by focussing on factors which strongly influence the likelihood of fire, in part-

icular fuel load (e.g. Sannier *et al* 2002) or canopy water content (Ustin *et al* 1998). For example, in the Etosha National Park, Namibia the potential for fires starting and propagating is strongly linked to the amount of biomass, so biomass maps, derived from satellite data, present a useful management resource (Sannier *et al* 2002).

Vegetation stress detection

Vegetation stress can be monitored through standard vegetation indices, via shifts in the red edge or by estimating canopy biophysical properties such as canopy water content (Ceccato *et al* 2001) or chlorophyll content (Jago *et al* 1999). The red edge is a sharp increase in reflectance in vegetation spectra which occurs between 670 760 nm. It marks the transition between low red and high NIR vegetation reflectance. Variations in the position of the red–edge, especially the inflection point, have been shown to indicate plant status (Guyot *et al* 1992). The high spectral resolution of current airborne imaging spectrometers enables shifts in the red edge to be measured (Dawson & Curran 1998).

Forest biodiversity

Remote sensing observations are not directly linked to forest biodiversity, although they can provide valuable information for forest biodiversity studies. Two principal methods have been used to analyse remote sensing data with regard to biodiversity. First, remote sensing data maybe used to produce vegetation maps, with field measurements of species distribution or abundance being used to relate specific vegetation classes to other species associated with the vegetation type (e.g. Fuller *et al* 1998). Second, the data maybe used to analyse forest structure, in particular, to quantify the level of fragmentation, presence of habitat corridors and the structure of forest edges (sharp or transitional edges) (Innes & Koch 1998; Hansen *et al* 2001). Both methods are based on the basic assumption that vegetation type and forest spatial structures influence the forest biodiversity and the survival of different species. The two methods are also similar in that they combine the wide coverage of the remote sensing data, with the more limited sampling of field observations, to provide a wider scale of coverage than field observations alone.

Forest temporal change

To some extent all the forest characteristics, which are observable with remote sensing data, can be measured over time to monitor forest change. Scene reflectance is influenced by canopy composition, which changes over time as vegetation growth cycles are often seasonally cyclic and typically species specific, so temporal signatures are particularly useful for vegetation analysis (Gerstl 1990; Achard & Estreugil 1995), phenological studies (Duchemin *et al* 1999) and discrimination between deciduous and coniferous species. Forest reflectance signatures may also experience enforced temporal change due to changes in environmental conditions such as climate change (Zhou *et al* 2001) and pollution events (Levesque & King 1999).

Conclusion

It is important that the constraints and limitations of remote sensing and remote sensing data are acknowledged so that the application of remote sensing data or derived products is carried out in an appropriate manner. The key limitation of remote sensing, with regard to forest science, is that the canopy is observed from above so that when dense canopies with complete (or near complete) canopy closure are encountered remote sensing observations saturate and lose their sensitivity to changes in the canopy. Despite this limitation remote sensing is an invaluable tool, especially with the global vegetation products now being routinely produced from MODIS and MISR observations. It remains the only practical means to scaling up forest properties, collected in the field or canopy, from local to regional or global scales.

Contact details

Dr Clare S. Rowland and Dr Terence P. Dawson, Environmental Change Institute, University of Oxford, 1A Mansfield Road, Oxford OX1 3SZ, UK.

Literature cited

Aber JD, Magill A, Boone R, Mellilo JM, Steudler P & Bowden R (1993). Plant and soil responses to chronic nitrogen additions at the Harvard Forest, Massachusetts. *Ecological Applications,* 3: 156–166

Aber JD (ed.), (1994). *Accelerated Canopy Chemistry Program Final Report to NASA-EOS-IWG,* National Aeronautics and Space Administration, Washington, DC

Achard F & Estreugil C, (1995). Forest classification of south-east Asia using NOAA AVHRR data. *Remote Sensing of Environment,* 54: 198–208

Bonan GB (1991). Atmosphere-biosphere exchange of carbon dioxide in boreal forests. *Journal of Geophysical Research,* 96: 7301–7312

Ceccato P, Flasse S, Tarantola S, Jacquemoud S, Gregoire JM (2001). Detecting vegetation leaf water content using reflectance in the optical domain. *Remote Sensing of Environment,* 77: 22–33

Cropper Jr WP & Gholz HL (1993). Simulation of the carbon dynamics of a Florida slash pine plantation. Ecological Modelling, 66: 231–249

Curran PJ (1994). Attempts to drive ecosystem simulation models at local to regional scales. In Foody GM & Curran PJ (eds.). *Environmental Remote Sensing from Regional to Global Scales,* Wiley and Sons, Chichester, 149–166

Curran PJ & Kupiec J (1995). Imaging spectrometry: A new tool for ecology, In Danson FM & Plummer SE (eds.). *Advances in Environmental Remote Sensing,* Wiley and Sons, Chichester, pp. 71–88.

Dawson TP & Curran PJ (1998).A new technique for interpolating the reflectance red edge position. *International Journal of Remote Sensing,* 19: 2133–2139

Dawson TP, Curran PJ, North PRJ & Plummer SE (1999). The propagation of foliar biochemical absorption features in forest canopy reflectance: A theoretical analysis. *Remote Sensing of Environment,* 67: 147–159

Duchemin B, Goubier J, Courrier G (1999). Monitoring phenological key stages and cycle duration of temperate deciduous forest ecosystems with NOAA-AVHRR data. *Remote Sensing of Environment,* 67: 68–82

Fuller RM, Groom GB, Mugisha S, Ipulet P, Pomeroy D, Katende A, Bailey R, Ogutu-Ohwayo R (1998). The integration of field survey and remote sensing for biodiversity assessment: a case study in the tropical forests and wetlands of Sango bay, Uganda. *Biological Conservation,* 86: 379–391

Gerstl SAW (1990). Physics concepts of optical and radar reflectance signatures. A summary review. *International Journal of Remote Sensing,* 11: 1109–1117

Gopal S & Woodcock C (1996). Remote sensing of forest change using artificial neural networks. *IEEE Transactions on Geoscience and Remote Sensing,* 34: 398–404

Graetz RD (1990). Remote sensing of terrestrial ecosystem structure: An ecologists pragmatic view. In Hobbs RJ & Mooney HA (eds.). *Ecological Studies: Remote Sensing of Biosphere Functioning*, Springer-Verlag, New York, pp. 5–31

Guyot G, Baret F & Jacquemoud S (1992). Imaging spectroscopy for vegetation studies. In Toselli F & Bodechtel J (eds.). *Imaging Spectroscopy: Fundamentals and Prospective Applications*, Kluwer Academic Publications, Dordrecht, pp. 145–165

Hall FG, Townshend JR & Engman ET (1995a). Status of remote sensing algorithms for estimation of land surface state parameters. *Remote Sensing of Environment*, 51: 138–156

Hansen MJ, Franklin SE, Woudsma CG & Peterson M (2001). Caribou habitat mapping and fragmentation analysis using Landsat MSS, TM, and GIS data in the North Columbia Mountains, British Columbia, Canada. *Remote Sensing of Environment*, 77: 50–65

Innes JL & Koch B (1998). Forest biodiversity and its assessment by remote sensing. *Global Ecology and Biogeography Letters*, 7: 397-419

Jago RA, Cutler MEJ & Curran PJ (1999). Estimating canopy chlorophyll concentration from field and airborne spectra. *Remote Sensing of Environment*, 68: 217–224

Levesque J & King DJ (1999). Airborne digital camera image semivariance for evaluation of forest structural damage at an acid mine site. *Remote Sensing of Environment*, 68: 112–124

Mayaux P, De Grande G, Malingreau J-P (2000). Central African forest cover revisited: a multi-satellite analysis. *Remote Sensing of Environment*, 71: 183–196

Miller GT (1992). *Living in the environment :An introduction to environmental science*. Wadsworth Publishing Co., California, 7th ed.

Peterson DL, Spanner MA, Running SW & Teuber KB (1987). Relationship of thematic simulator data to leaf area index of temperate coniferous forests. *Remote Sensing of Environment*, 22: 323–341

Plummer SE (2000). Perspectives on combining ecological process models and remotely sensed data. *Ecological Modelling*, 129: 169–186

Running SW & Coughlan JC (1988). A general model of forest ecosystem processes for regional applications, I. Hydrologic balance, canopy gas exchange and primary production processes. *Ecological Applications*, 42: 125–154

Sandmeier St & Deering DW (1999). Structure analysis and classification of boreal forests using airborne hyperspectral BRDF data from ASAS. *Remote Sensing of Environment*, 65: 280–291

Sannier CAD, Taylor JC & Plessis W Du (2002). Real-time monitoring of vegetation biomass with NOAA-AVHRR is Etoshu National Park, Namibia, for fire risk assessment. *International Journal of Remote Sensing*, 23: 71–90

Sellers PJ, Los SO, Tucker CJ, Justice CO, Dazlich DA, Collatz GJ & Randall DA (1996). A revised land surface parameterization (SiB2) for atmospheric GCMs. Part II: The generation of global fields of terrestrial biophysical parameters from satellite data. *Journal of Climate*, 9: 706–737

Sellers PJ, Dickinson RE, Randall DA, Betts AK, Hall FG, Berry JA, Collatz GJ, Denning AS, Mooney HA, Nobre CA, Sato N, Field CB, Henderson-Sellers A (1997). Modeling the exchanges of energy, water and carbon between continents and the atmosphere. *Science*, 275: 502–509

Treitz P & Howarth P (2000). Integrating spectral, spatial, and terrain variables for forest ecosystem classification. *Remote Sensing of Environment*, 72: 268–289

Ustin SL, Roberts DA, Pinzon J, Jacquemoud S, Gardner M, Scheer G, Castageda CM & Palacios-Orueta A (1998). Estimating canopy water content of chapparral shrubs using optical methods. *Remote Sensing of Environment*, 65: 280–291

Zhou L, Tucker CJ, Kaufman RK, Slayback D, Shabanov NV, Myneni RB (2001). Variations in northern vegetation activity inferred from satellite data of vegetation index during 1981 to 1999. *Journal of Geophysical Research-Atmosphers*, 106: 20069–20083

Satellite Imaging of Forest Canopies

By
H Balzter and
F Gerard

SINCE the launch of the first Landsat satellite in 1972 satellite remote sensing has become widely used in the observation of forest canopies. Its particular value is the ability to deliver repeated spatially comprehensive datasets of large areas. Because many forests are not easily accessible, remote sensing often complements other measurement techniques.

Optical sensors

Optical sensors record the proportion of incident solar radiation reflected from targets on the Earth's surface in the visible (0.4 mm–0.7 mm), near infrared (0.7 mm–1.3 mm) and short-wave infrared (1.4 mm–1.8 mm) spectrum. There are a wide variety of spaceborne optical sensors currently in use (see Table 1). The application of the data recorded by these sensors is determined by the sensors' characteristics with respect to spatial, spectral and temporal resolution and viewing geometry. Limitations to what was technically possible at the time when these sensors were constructed have resulted in compromises between these characteristics. For instance, sensors with a very fine spatial resolution (e.g. SPOT-HRV: 10 m, IKONOS: 1 m) tend to record a limited number of broadband channels and have a low repeat frequency. This type of data is very useful when infrequent but detailed spatial information is required for relatively small areas. Sensors such as VEGETATION or AVHRR have a much coarser resolution of 1 km but collect daily data. These data are often used to monitor seasonal or frequent and large scale events such as the onset of the growing season or wild fires. A major drawback of optical remote sensing is that it is affected by variable atmospheric conditions and cannot see through clouds.

Different physical processes control the reflectance of a forest canopy at different spatial scales. Leaves and branches represent the smallest scattering elements. The reflectance, transmittance and absorption at that scale are governed by the elements' surface characteristics, cellular structure and chemical composition. At medium scales, the arrangement of the leaves and branches within the tree and shrub canopies is a distinct factor. The multiple scattering occurring between the elements at this scale causes the forest canopy to reflect in an anisotropic manner. Larger scale effects depend on the spatial arrangement of individual trees and shrubs and the distribution of forested areas in a heterogeneous landscape. At the larger scale the forest reflectance may also be affected by reflectances from other surface features present in the heterogeneous landscape (e.g. bare rock, man made features, soil, water).

Optical remote sensing can provide information on forest extent, regeneration stage, forest type, phenology and to some degree forest canopy cover, leaf area index and biomass.

Forest extent

The use of satellite imagery has improved and accelerated the global and continental mapping of forested/non-forested areas. It provides data, which can be

Table 1 Spatial resolution and spectral range of different satellites

SENSOR (SATELLITE)		SPATIAL RESOLUTION	SWATH (KM)	SPECTRAL RANGE (µM)	REPEAT FREQ. (DAYS)	ACQUISITION DATE
AVHRR (NOAA – POES series)	Andvanced Very High Resolution Radiometer	1.2 km	2399	0.58–12.4	1	From 1978
MSS (Landsat 1 to 5)	Multi-spectral Scanning System	80 m	185	0.5–1.1	18, 16	From 1972
TM (Landsat 4 to 5)	Thematic Mapper	30 m	185	0.45–12.5	16	From 1982
HRV & HRVIR (SPOT 1 to 4)	High Resolution Visible and Infrared	10 m	117	0.50–0.89	26	From 1986
LISS-II (IRS)	Linear Imaging Self Scanning System	36 m	74	0.46–0.86	22	From 1994
ETM (Landsat 7)	Enhanced Thematic Mapper	30 m	183	0.45–12.5	16	From 1999
VEGETATION (SPOT)		1 km	2200	0.61–1.75	1	From 1998
ATSR series (ERS-2, ENVISAT)	Along Track Scanning Radiometer	1 km	500	0.55–12	3	From 1995
IKONOS		1 m, 4 m	n.a.	0.45–0.89	n.a.	From 1999
MODIS (TERRA)	Moderate Resolution Imaging Spectro-radiometer	250 m to 1 km	2330	0.4–14.4	2	From 2001
MERIS (Envisat)	Medium Resolution Imaging Spectrometer	300 m to 1.2 km	1450	0.4–1.05	2	From 2002

interpreted consistently across regions, countries and continents. In 1979 the FAO/UNEP initiated a project to update baseline information on the tropical forests (Lanly 1982). The method used combined information from remotely sensed data (e.g. Landsat MSS and TM and airborne SAR) with data acquired from conventional inventory techniques. When this exercise was repeated in the early nineties, as part of the TREES project of the European Union, the estimations were mainly based on AVHRR imagery combined with existing data from archives. The Brazilian Space Agency has been using Landsat TM to monitor deforestation in Brazil for some years now (Tardin and daCunha 1990) whilst the NASA Landsat Pathfinder project set out to estimate tropical deforestation over the past 20 years across Amazonia, Central Africa, and South-East Asia by analysis of Landsat MSS and TM imagery (Chomentowski et al 1994). More recently an AVHRR-based forest proportion map of Europe was produced through a collaborative effort between the University of Helsinki (Finland), Stora Enso Forest Consulting Ltd. (Finland) and the Joint Research Centre (Italy) (Hame et al 2001). A mosaic of 49 AVHRR scenes from 1996 and 1997 were used to cover a total of 12 countries. The attraction of using AVHRR for mapping global or continental forest extent is that it is freely available, has a wide 2399 km swathe width, thus covering large areas at one time, and revisits the same area every day, increasing the chance of cloud free coverage considerably. The main problem with coarse spatial resolution imagery (1 km) such as AVHRR is the general tendency of forest area underestimation if the forest cover is fragmented and overestimation if the forest cover is continuous over large areas. Although techniques have been developed to improve forest area estimates for highly fragmented landscapes which involve for example the integration of AVHRR with 30 m TM imagery, a significant improvement in area estimates will come from the medium spatial resolution imagery of MODIS (available since 2000) and MERIS (available in summer 2002) and their derived land cover products (http:/eosdatainfo.nasa.gov/eosdata/terra/modis).

Regeneration

Optical spectral signatures of forests are highly responsive to different stages of regeneration and canopy closure. Work by Steiniger (1996), using Landsat™ data, suggested that it is possible to distinguish tropical primary forests from secondary forests of up to 14 years of age on the basis of their spectral response. Alves and Skole (1996) used time-series of SPOT-HRV images to analyse secondary forest regrowth in an area in Rondonia, Brazil from 1986 to 1992. The observed differences in canopy reflectance in the red and infrared spectrum through the successive stages appear to be due mainly to changes in canopy geometry, which result in changes in the patterns of shadow thrown by the canopy. The short wave infrared (SWIR) is particularly sensitive to water absorption and has shown to be critical to the mapping of regenerating fire scars in the boreal forests of Canada which have less vegetation water content then the surrounding unburned forests.

Forest type

Most attempts to map the different forest types using remotely-sensed data have not gone beyond the national scale and have reported varying success. Efforts to map vegetation in greater detail at global or continental scales with remotely sensed data are few. One example is the IGBP-DIS Global Land Cover Project which resulted in the successful compilation of a global digital land cover dataset at 1 km resolution (Eidenshink and Faundeen 1994). It identifies a few broadly defined forest types that have functional significance for global modelling (e.g. leaf-shape and leaf-seasonality).

An ideal forest inventory would have to adopt a classification system which has relevance for applications in ecology, forest management and environmental research. However this may not be possible. Any vegetation classification system is the result of a compromise between its scope of use, its level of detail and its spatial application. As a result, a wide variety of forest classification systems exist, generating problems of consistency and comparability. The use of remote sensing to map forest types introduces another dimension to forest classification. With remotely sensed data, forests are classed according to their spectral characteristics.

In the temperate zone, there has been some success in differentiating homogeneous stands of broad-leaved and needle-leaved species from satellite imagery. Tree species show different spectral characteristics caused by differences in leaf, branch and stem spectral properties. Nevertheless, forest species identification in temperate and tropical forests with high species diversity has had limited success. The generation of forest maps, which aim at classifying forests from a floristic

point of view, may therefore not always be possible from remotely sensed information alone. On the other hand, differences in structural patterns and canopy physiognomy identified from aerial photographs frequently coincide with differences in floristic composition. Moreover it has been suggested that physiognomy may be the most important class attribute that influences the spectral response of vegetation. So, the creation of forest type maps based on physiognomic or structural characteristics may gain from using remotely sensed information. Nevertheless many different factors determine the interaction between radiation and forests and understanding the relationship between forest reflectance and the traditional physiognomic, floristic and ecological classification parameters is essential if we are to produce useable forest type maps from remotely sensed data (Malingreau *et al* 1992).

Phenology

The detection and accurate measurement of phenological events could, help for example:
- identify isolated climatic or stress events such as drought, flooding, disease,
- provide evidence of climate change,
- identify areas susceptible to fire or
- determine different types of forests and vegetation.

Leaves are major contributors to the spectral signal of vegetation. At red wavelengths (0.6–0.7 mm) leaf chlorophyll will strongly absorb light, which provides the energy needed for photosynthesis. In the near infrared region (0.7–0.8 μm), light undergoes strong scattering in the intercellular leaf cavities. Large-scale seasonal events characterised by synchronised leaf-flush, leaf-senescing or leaf-fall, are therefore likely to change the spectral signature of a canopy. Vegetation indices (VIs), such as the Normalised Difference Vegetation Index (NDVI), have been developed to enhance these spectral responses. Most of them are ratio or linear combinations of reflectances in the red and near infrared part of the spectrum and vary therefore with phenology.

Time series of the NDVI calculated from AVHRR imagery, have been used by the remote sensing community to detect seasonal and inter-annual phenology patterns at the biomes scale (e.g. Malingreau *et al* 1996). The NDVI and other VIs have also been used to detect drought patterns (Liu & Kogan 1996) and measure vegetation growth condition (Curran 1980). More recently Duchemin *et al* (1999) developed a method using AVHRR NDVI time series for monitoring the phenological cycle of French deciduous forests over a six year period. The NDVI derived phenology revealed differences with respect to forest species composition and local climate. However, VIs have shown sensitivity to non-vegetation related factors such as soil background reflectance, atmosphere and solar angle, and exhibit saturation at canopy closure. The use and interpretation of NDVI time series and other VIs should therefore be treated with care.

Canopy cover, leaf area index and biomass

A number of studies have explored the relationship between vegetation reflectance and canopy biophysical properties. Canopy cover is the most strongly correlated with canopy reflectance in the red spectrum. However, the relationship between red reflectance and canopy cover is affected by solar zenith angle, shadow and the type of background and becomes less significant at high canopy covers (from 65%).

Estimating forest leaf area index (LAI) from remotely sensed data has been the subject of many studies over the last two decades. Semi-empirical relationships, between VIs and LAI have had varying success. The main difficulty associated with the use of most of these VIs is their sensitivity to soil type and moisture before canopy closure and shadow after canopy closure (Huete 1987). Another important problem is saturation of most VIs of forest canopies at LAI values greater than 3–4.

The estimation of forest biomass is still a major challenge because of the structural complexity and variability of these forests and the limited amount of reliable biomass (above- and below ground) information available. Several authors have investigated the potential of remotely sensed data to derive forest volume or above-ground biomass from semi-empirical equations (e.g. Roy & Ravan 1996, Trotter *et al* 1997). Some assumptions used when correlating spectral reflectance with above-ground biomass are:
- timber volume increases with age and tree size, both of which have an effect on the shadow content of the imagery.
- above-ground biomass is related to forest type, age and stem density parameters which jointly affect forest reflectance values.

Others have looked at ways of calculating total biomass or biomass increments indirectly from semi-empirical estimates of LAI or APAR (Absorbed Photosynthetically Active Radiation) using physiologically-based ecosystem models (e.g. Verstraete *et al* 1996, Coops *et al* 1998). A major drawback of using semi-empirical relationships is the requirement for re-calibration or re-formulation of the model for areas outside the ones studied and for different times of year. Moreover, remotely sensed reflectance values from forest canopies are the result of complex light-canopy interactions, soil reflectance and atmospheric effects, all of which will introduce a high level of variability and uncertainty to semi-empirical relationships. The development of light-canopy interaction models (e.g. North 1996) offers a physically based and more flexible alternative to semi-empirical equations. Inversion of such models may, in the long term, enable more reliable estimation of forest cover, LAI and biomass.

Radar sensors
Synthetic Aperture Radar (SAR) is an active sensor transmitting pulses of horizontally or vertically polarized electromagnetic waves in the microwave spectrum (1–150 cm, typically 3–25 cm) and receiving back the backscattered radiation. A sensor is characterised by the wavelength, polarization, range and azimuth resolution. Operational radar wavelengths used for forest mapping are X-band (3.1 or 3.5 cm wavelength), C-band (5.65 cm), L-band (24 cm) and P-band (30–60 cm). Important satellite features are the available swath widths and the repeat cycle. More advanced SAR techniques exploit multiple polarizations (orientations of the electromagnetic fields), multitemporal images, and interferometric image pairs. SAR interferometry uses two SAR images taken from similar viewing positions with a short time-lag. Interference between the two signals can be used to compute the interferometric coherence (correlation) and the interferometric phase which

contain information on the three-dimensional structure of the scattering elements in the imaged area.

In forests, the radiation penetrates the top vegetation layer to a certain depth and is scattered by stems, branches, twigs, leaves or needles. Longer wavelengths tend to penetrate deeper into the vegetation canopy, so that multi-wavelength imaging can provide information on canopy density. Microwaves can penetrate clouds, dry snow and to some extent rain, and are independent of daylight which is important for imaging during the boreal winter darkness.

SAR sensors can be used to provide estimates of parameters such as forest area, forest thermal state (frozen/thawed), forest biomass density and tree height (Balzter 2001, Kasischke 1997). The magnitude of microwave backscatter is a result of the geometric and dielectric properties of the surfaces or volumes imaged. It is sensitive to the viewing and surface geometry (look angle, topography), surface roughness (surface slope, variation of surface height, plant geometry) and water content of surface materials (crop and soil moisture, snow wetness).

Early radar techniques were developed using airborne and space shuttle based SAR sensors (NASA's Shuttle Imaging Radar missions). Research satellites carrying SAR sensors include the C-band instruments on board the European ERS-1 (launched 1991), ERS-2 (1995) and ENVISAT (to be launched 2002), the Canadian RADARSAT series and the L-band instruments on the Japanese JERS-1 (1992) with the follow-on mission ALOS planned to be launched in 2004. Despite the emphasis of these missions on research and development, a large number of applications have exploited the data. Some examples are presented below.

Applications
Forest cover map of Siberia

The boreal forest belt of Russia is a significant carbon pool. Because of the lack of infrastructure and the remoteness of Siberia, forest inventories are not carried out frequently enough to provide timely information on the boreal ecosystem. Due to a gap in the global network of satellite receiving stations, Siberia had not been covered by radar before. In 1997/1998, the German Aerospace Research Centre (DLR) deployed a mobile receiving station in Mongolia to record SAR signals from the Japanese JERS-1 and the two European ERS satellites (Schmullius 1997). In the European Commission-funded SIBERIA project (http://pipeline.swan.ac.uk/siberia/) 600 JERS-1 SAR images and 356 ERS-1 and ERS-2 SAR images were processed. The JERS-1 SAR operates at 24 cm wavelength, horizontal transmit and receive mode (HH), while the ERS-1 and ERS-2 radars work at 5.6 cm wavelength and vertical transmit and receive mode (VV). ERS-1 and ERS-2 were operated by the European Space Agency to image the same areas from similar viewing positions with one-day time-lag. ERS Tandem SAR interferometry provided:
- Coherence measurements, quantifying the correlation between the two signals, and
- Phase measurements used to derive topographic height. All satellite images were (i) multi-looked, (ii) resampled to a pixel spacing of 50 m, (iii) coregistered, (iv) speckle filtered, (v) masked for high topography, and, (vi) mosaicked.

Russian forest inventory data were used to develop an adaptive image classification algorithm based on JERS-1 radar backscatter and ERS interferometric coherence. Total growing stock volume [m³/ha] is the target parameter. It is correlated to the number and size distribution of scattering elements in the forest canopy, and thus to the radar signal. Following explorative studies, the classes 'Water', 'Smooth open areas' (including bogs, agriculture and grassland) and four total growing stock volume classes, '£20 m³/ha', '20–50 m³/ha', '50–80 m³/ha' and '>80 m³/ha' were defined as target classes. The lowest total growing stock class includes tundra, fire scars, shrublands and clearings with birch regrowth. The two intermediate classes represent stages of secondary regrowth, and the highest class shows potentially economically important forest stands. Empirical models with image-specific parameters accounted for image-to-image variation, and a new contextual classification algorithm, the Iterated Contextual Probability (ICP) algorithm, was used to eliminate isolated misclassified pixels. The entire classification process was automated on a UNIX workstation. Reclassifying the whole map would only take 24 hrs. The provided methods can potentially be used in future forest mapping projects. First tests on sites in Brazil, the UK and Finland showed promising accuracies between 70% and 83% (Tansey *et al*, in prep.).

The entire classified Siberian forest cover map is shown in Figure 1, and one of the map sheets in Figure 2. The map accuracy was assessed by an independent ground survey and the coefficient of agreement κ_w is high for the seven test territories (Balzter *et al*, in prep.). It varies between 0.73 and 0.97 (pooled $\kappa_w = 0.94$). The user accuracies for each individual class are all greater than 80%. For the Russian forest enterprises in Krasnoyarsk Kray and Irkutsk Oblast the map will provide a timely update of their forest inventory databases.

The map fills a gap in the data coverage by the Global Forest Monitoring Programme led by the Japanese Space Agency (NASDA) in collaboration with the German Aerospace Center DLR, the European Space Agency ESA and the

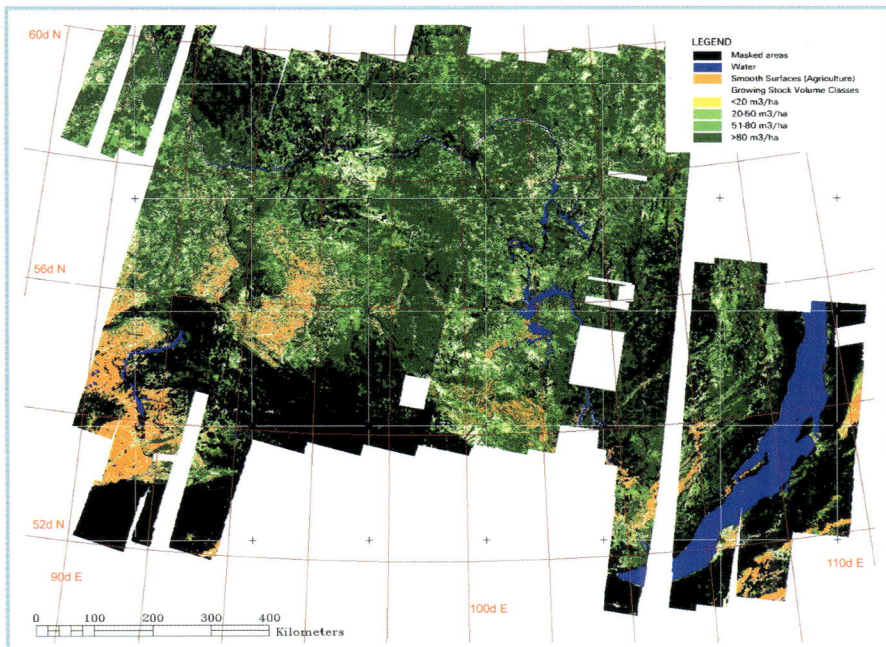

Figure 1 Overview of the Forest Cover Map, a mosaic of classified SAR images of Central Siberia. EC ENV4-CT97-0743-SIBERIA, ESA 97/98, NASDA GBFM, DLR.

European Commission's Joint Research Centre. It complements other SAR mosaics of forests, like the Equatorial Africa, Amazon and southeast Asia JERS-1 mosaics (http://southport.jpl.nasa.gov/).

Deforestation mapping in the Amazon basin

Forested areas in the Brazilian Amazon are undergoing rapid changes as a result of human influence. In the Rondônia State, for instance, land exploitation was increased by the construction in 1968 of the Cuiabá – Pôrto Velho road. Tropical forests are exploited for timber extraction, shifting cultivation and permanent agriculture and pasture. The large-scale conversion of forest to agricultural fields and pasture leads to biodiversity loss and changes in carbon storage in forest biomass and soil.

ERS SAR interferometry using ERS-1/ERS-2 Tandem data has been applied to mapping deforestation in the Rondônia State in the Brazilian Amazon by (Strozzi, 1999). A hierarchical classification of urban areas, water, forest and open land was carried out in order to quantify the extent of deforestation near the city of Ariquemes. The classification approach is based on thresholds for coherence, backscatter and backscatter change images.

A Werner-Wegmüller composite of coherence (red), backscatter (green) and temporal change of backscatter (blue) is shown in (Figure 3). In the figure forest appears green because of low coherence, medium backscatter and low backscatter change. The yellow color of the city of Ariquemes is the result of high coherence, high backscatter intensity and low backscatter change. The blue colour of water is caused by high temporal change of backscatter combined with low coherence and low backscatter. The road from Pôrto Velho to Cuiabá follows the Rio Jameri and the city of Ariquemes is located in the center of the deforested area. Deforested areas show a red colour because their coherence is higher than over forest.

Interferometric SAR at C-band can be used to automatically classify image

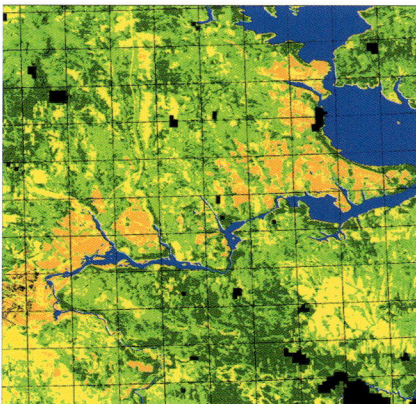

Figure 2 Example of the 123 map sheets showing the forest classification from SAR. Scale 1:200,000; 10 km grid lines. EC ENV4-CT97-0743-SIBERIA, ESA 97/98, NASDA GBFM, DLR.

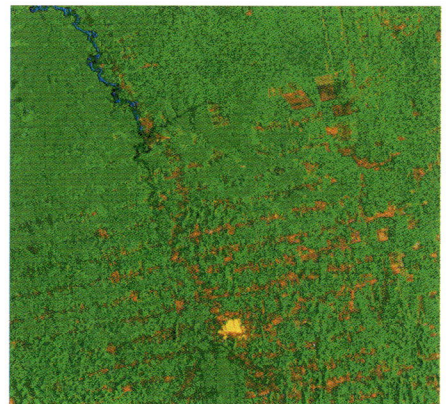

Figure 3 Deforestation map from ERS SAR interferometry over Rondônia in the Brazilian Amazon. Colour composite of coherence (red), backscatter (green) and the temporal change of backscatter (blue). 1999 ESA. ERS data courtesy of AO 3-178.

mosaics detecting deforestation because of the high coherence for reduced vegetation density. The rain forest shows the typical strong fragmentation known from optical imaging (Rignot 1997). Large parcels of deforested land are mainly used for ranching, while small blocks are mainly used for agriculture. Some of these parcels have been abandoned and are in different stages of secondary regeneration.

Canopy height at a British test site

As part of the SAR and Hyperspectral Airborne Campaign 2000 in the UK, a study was conducted examining tree height estimation from single-pass SAR

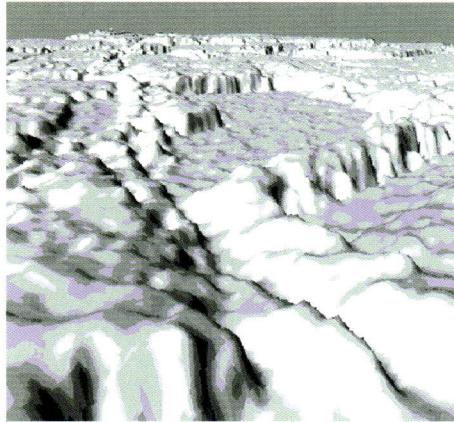

Figure 4 3D view of the canopy height map of Thetford forest, UK, derived from the difference of X-band SAR interferometry and the OS DTM. Rows of trees and height differences between forest stands are visible. 2001 NERC. Data courtesy of NERC/BNSC

interferometry at X-band using images acquired by the German airborne E-SAR. The concept is based on the extraction of a Digital Surface Model (DSM) from interferometric SAR image pairs. Because of the dielectric properties of live phytomass the interferometric height represents the sum of topographic and vegetation canopy height. To map canopy height a Digital Terrain Model (DTM) is subtracted from the DSM. The DTM can be an external data source or an interpolated product from the DSM based on vegetation-free areas and a suitable algorithm (like bilinear interpolation, triangulation or kriging). The accuracy of the canopy height map depends both on the quality of the interferometric SAR data products and the quality of the DTM. Here, an external Ordnance Survey (OS) DTM derived from digitised contour lines (10 m pixel spacing) was used. Figure 4 shows a 3D representation of tree height estimated from the X-VV and OS DTMs. Boundaries between even-aged stands and a line of trees along a bridleway are clearly visible. The precision of the estimated tree height was assessed using tree height values derived from species, stand age and yield class in the Forestry Commission GIS database and published general yield class curves (Forestry Commission 1981). A problem is the underestimation of canopy height for older stands (>20 m) in Figure 5a. These stands have a large gap fraction and the SAR signal is a mixture of canopy and ground contributions. A simple gap fraction model based on allometric data was used together with a correction for the incidence angle effect (Balzter 2001). Figure 5b shows the results of this correction.

This technique is so far only available from airborne platforms and shuttle radar data such as the Shuttle Radar Topography Mission. In the future there may be X-band interferometric systems on board of satellites. Other wavelengths and lower resolutions are available from satellite (C-band from ERS Tandem mission), but not operationally and only in repeat-pass mode which causes a higher height error.

Fire scars in Canada

Forest fires play a major role in the terrestrial carbon cycle. Global climate change is thought to alter the frequency, size and severity of fire events by changing the length of the fire season, the vegetation moisture content and fire risk. Monitoring disturbance in the forest biome is vital for understanding forest dynamics and predicting potential impacts of global environmental change. The standard methods of fire observation are individual wildfire reports prepared by the forest fire management agencies. In Canada, wildfire reports typically comprise data on fire number, location (point of ignition), size and start and end date. The burned area is derived either from aerial sketch maps or a posteriori aerial photographic interpretation. This can be supplemented by information on fire type and severity derived indirectly from height of burn-marks on stems, fire spread rate, flame height, level of combusted litter or mortality of trees. These indicators are used to define a particular fire type (on-ground, crown) into three intensity levels – low, medium and high. In Canada, a major initiative was started to co-ordinate,

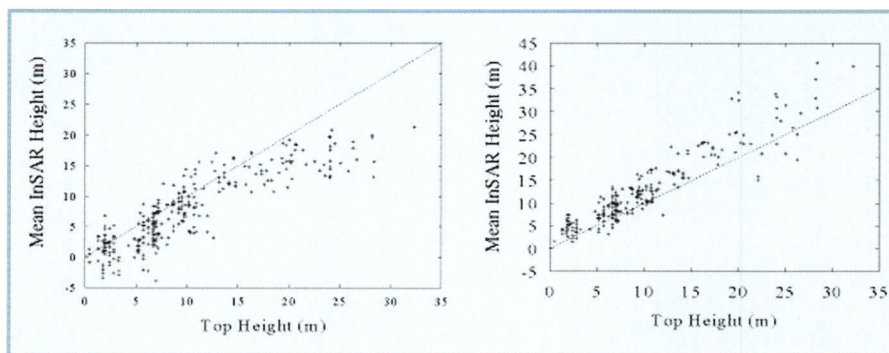

Figure 5 Mean canopy height per forest stand estimated from the InSAR imagery with the OS DTM compared to top height from the yield class model. (a) Uncorrected values, for stands > 20 m the stand height is underestimated. (b) Corrected values for gap fraction and incidence angle show an improved quality of the product. 2001 ESA.

compile and rationalise wildfire reports from different provinces into a comprehensive annual database focusing on those fires greater than 200 hectares. These fires represent 98% of the total area burned.

The Canadian Large Fire Database (LFDB) represents a compilation, co-ordinated by the Canadian Forest Service, of all fires >200 ha which have occurred in Canada since the 1950s (www.nofc.forestry.ca/fire/frn/English/ClimateChange/LFDB8095.htm). The dataset includes digitized and geo-referenced maps of final fire perimeters. The LFDB is now being extended backwards in time to include the full archive from each of the provinces and territories. During a study in the SPOT-VEGETATION Preparatory Programme (Plummer *et al* 2002, Ferreruela *et al* in press) remote sensing methods for detecting regenerating fire scars using the SPOT-VEGETATION sensor were developed. The images were divided into homogeneous polygons by a segmentation algorithm and combined to produce the calculated probability of each segment representing a fire scar. Figure 6 shows part of a SPOT-VEGETATION image, the result of the segmentation algorithm and the fire probability map. These fire scar maps will add spatial information to the existing fire data.

Acknowledgements

The SIBERIA project was partly funded by Framework 4 of the European Commission, Environment and Climate, Area 3.3: Center for Earth Observation, Theme 3: Space Techniques Applied to Environmental Monitoring (Contract No. ENV4-CT97-0743-SIBERIA). SAR data were provided by ESA's 3rd ERS Announcement of Opportunity (Project Number AO3.120 (SIBERIA) and NASDA's JERS initiative "Global Boreal Forest Mapping". The satellite data were received by a mobile receiving station of the German Remote Sensing Data Center of DLR (DFD) at Ulaanbaatar, Mongolia. We wish to thank the other members of the SIBERIA team, particularly Prof. Chris Schmullius for making it happen. The ERS SAR data of Brazil are courtesy of ESA and were provided under AO 3-178. Tazio Strozzi (Gamma Remote Sensing) is acknowledged for producing the Brazilian image classification. The SAR and Hyperspectral Campaign 2000 (SHAC) was sponsored by the British National Space Centre and the UK Natural Environment Research Council. SPOT images were provided by the Centre National d'Études Spatiales, France, contract no. 95/CNES/0406.

Contact details

Dr Heiko Balzter, Head of Biophysical Modelling Group, Section for Earth Observation, Centre for Ecology and Hydrology (CEH), Monks Wood, Abbots Ripton, Huntingdon, Cambridgeshire PE28 2LS, UK. Email: hbal@ceh.ac.uk

Literature cited

Achard F & Blasco F (1990). Analysis of vegetation seasonal evolution and mapping of forest cover in west Africa with the use of NOAA AVHRR HRPT data. *Photogrammetric Engineering and Remote Sensing*, 56(10), 1359–1365.

Alves DS & Skole DL (1996). Characterising land cover dynamics using multi-temporal imagery. *International Journal of Remote Sensing*, 17(4), 835–839.

Balzter H (2001). A review of remote sensing of forests with interferometric Synthetic Aperture Radar (InSAR) techniques. *Progress in Physical Geography*, 25: 159–177.

Balzter H, Saich P, Luckman AJ, Skinner L & Grant J (2001). Forest stand structure from airborne polarimetric InSAR. *3rd International Symposium on Retrieval of Bio- and Geophysical Parameters from SAR Data for Land Applications (LANDSAR 2001)*, Sheffield, European Space Agency.

Balzter H, Talmon E, Wagner W, Gaveau D, Plummer S, Yu JJ, Quegan S, Davidson M, Le Toan T, Gluck M, Shvidenko A, Tansey K, Luckman A & Schmullius C, in prep., *Accuracy assessment of a large-scale forest map of Central Siberia from Synthetic Aperture Radar.*

Chomentowski WH, Salas WA & Skole DL (1994). Landsat Pathfinder project advances defor-

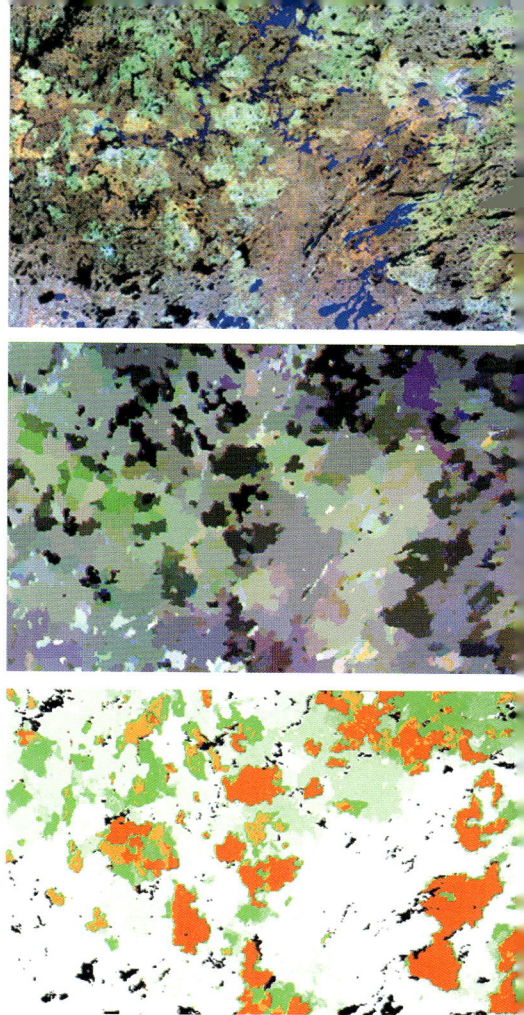

Figure 6 Top: false colour composite of SPOT-VGT image for an area of 942 km x 569 km in Thompson Manitoba, Canada; Middle: three segmentation results combined; Bottom: the fire scar probability map produced from the image segmentations, red indicates a high probability. 2001 NERC.

estation mapping. *GIS World*, 7(34–38.

Coops NC, Waring RH & Landsberg JJ (1998). Assessing forest productivity in Australia and New Zealand using a physiologically-based model driven with averaged monthly weather data and satellite-derived estimates of canopy photosynthetic capacity. *Forest Ecology and Management*, 104(1–3), 113–127.

Curran P (1980). Multispectral remote sensing of vegetation amount. *Progress in Physical Geography*, 4(315–341.

Duchemin B, Goubier J & Courrier G (1999). Monitoring phenological key stages and cycle duration of temperate deciduous forest ecosystems with NOAA/AVHRR data. *Remote Sensing of Environment*, 67(68–82.

Eidenshink JC & Faundeen JL (1994). The 1km AVHRR global land data set – 1st stages in implementation. *International Journal of Remote Sensing*, 15(17), 3443–3462.

Ferreruela A, Wadsworth R, Gerard F & Plummer S (in press). Mapping 'Historical' Fire Scars in the Canadian Boreal Forests. *GIS Research UK: 10th Annual Conference*, Sheffield, 3–5 April 2002.

Forestry Commission (1981). *Yield models for forest management* (booklet 48). Forestry Commission, Edinburgh,

Hame T, Stenberg P, Andersson K, Rauste Y, Kennedy P, Folving S & Sarkeala J (2001). AVHRR-based forest proportion map of the Pan-European area. *Remote Sensing of Environment*, 77(76–91.

Huete AR (1987). Soil and sun angle interactions on partial canopy spectra. *International Journal of Remote Sensing*, 8(1307–1317.

Kasischke ES, Melack JM & Dobson MC (1997). The use of imaging radars for ecological applications – A review. *Remote Sensing of Environment*, 59: 141–156.

Liu WT & Kogan FN (1996). Monitoring regional drought using the Vegetation Condition Index. *International Journal of Remote Sensing*, 17(14), 2761–2782.

Malingreau JP, Verstraete MM & Achard F (1992). Monitoring global tropical deforestation: a challenge for remote sensing. *TERRA-1: Understanding the terrestrial environment: The role of earth observation from space*, 121–131,Taylor & Francis.

Malysheva N, Shvidenko A, Nilsson S, Petelina S & Oeskog A (2000). An overview of remote sensing in Russian forestry. *International Institute of Applied Systems Analysis*, Vienna, 83.

Myneni RB, Nemani RR & Running SW (1997). Estimation of global leaf area index and absorbed par using radiative transfer models. *IEEE Transactions On Geoscience and Remote Sensing*, 35: 1380–1393.

North (1996). Three-dimensional forest light interaction model using a Monte Carlo method. *IEEE Transanctions on Geoscience and Remote Sensing*, 34(4), 946–957.

Plummer S, Gerard F, Iliffe L, Wyatt B & Wadsworth R (2002). Fire scar detection with SPOT-VEGETATION data. *Final report for the SPOT-VEGETATION Preparatory Programme, Phase II*. Centre for Ecology and Hydrology Monks Wood, UK.

Rignot E, Salas WA & Skole DL (1997). Mapping deforestation and secondary growth in Rondonia, Brazil, using imaging radar and thematic mapper data. *Remote Sensing of Environment*, 59: 167–179.

Roberts DA, Nelson BW, Adams JB & Palmer F (1998). Spectral changes with leaf aging in Amazon caatinga. *Trees-Structure and Function*, 12(6), 315–325.

Roy PS & Ravan SA (1996). Biomass estimation using satellite remote sensing data – an investigation of possible approaches for natural forest. *Journal of Biosciences*, 21(4), 535–561.

Schmullius CC & Rosenqvist A (1997). Closing the Gap – a Siberian Boreal Forest Map with ERS-1/2 and JERS-1. *3rd ERS Symposium on Space at the Service of our Environment*, Florence, Italy, ESA.

Steiniger MK (1996). Tropical secondary forest regrowth in the Amazon: age, area and change estimation with Thematic Mapper data. *International Journal of Remote Sensing*, 17(1), 9–27.

Strozzi T, Wegmüller U, Luckman AJ & Balzter H (1999). Mapping deforestation in Amazon with ERS SAR interferometry. *IEEE International Geoscience and Remote Sensing Symposium IGARSS 99*, Hamburg, IEEE.

Tansey KJ, Luckman AJ, Skinner L, Balzter H, Strozzi T & Wagner W, in prep., *Classification of global forest volume resources using ERS tandem coherence and JERS intensity data.*

Tardin AT & daCunha RP (1990). *Evaluation of Deforestation in the Legal Amazon Using Landsat-TM Images.* INPE-5015-RPE/609, INPE.

Tomppo E (1995). Finnish National Forest Inventory. *Paperi Ja Puu-Paper and Timber,* 77: 374–378.

Tomppo E (1996). Application of remote sensing in Finnish National Forest Inventory. *Application of Remote Sensing in European Forest Monitoring,* Vienna, Joint Research Centre, European Commission, BOKU Joanneum Research.

Trotter CM, Dymond JR & Goulding CJ (1997). Estimation of timber volume in a coniferous plantation forest using Landsat TM. *International Journal of Remote Sensing,* 18(10), 2209–2223.

Verstraete F, Patyn J & Myneni RB (1996). Estimating net ecosystem exchange of carbon using the normalised difference vegetation index and an ecosystem model. *Remote Sensing of Environment,* 58(1), 115–130.

Wulder M (1998). Optical remote-sensing techniques for the assessment of forest inventory and biophysical parameters. *Progress in Physical Geography,* 22: 449–476.

Multi-spectral and Hyper-spectral Airborne and Satellite Technologies for Assessing the Forest Canopy

By
Arturo
Sánchez-Azofeifa

TECHNIQUES using multi-spectral airborne and satellite remote sensing to observe canopies have been developed since the early 1980's. The main limitations of current methods for accessing canopy information are restrictions associated with spatial and spectral resolution of the sensors (e.g. Landsat™ has 7 spectral bands and SPOT has only four). These restrictions impose significant limitations on the analysis of forest canopies. Current problems are being solved by using hyperspectral airborne and remote sensing techniques. Hyperspectral data sets generate 'data cubes' which are produced from 200+ bands rather than the 7 bands used with the Multispectral satellites (e.g. TM or IKONOS). There is as yet no clear agreement on standardized approaches for the interpretation of hyperspectral data. Most of the literature is still discussing innovative applications which deal with lineal and non-lineal techniques aimed to extract information on species differentiation or different biochemical properties of canopies (e.g.chlrophyl a and b).

Choosing a location

There are four main criteria used for site selection:
1. Accessibility: need to be able to bring field spectrometers to the site (instruments used to measure in-situ spectral reflectance from leaves, understory and canopies).
2. Species composition variability: sites are selected based on the Holdridge Complexity Index (HCI). The HCI is a function of the number of tree species present on a plot, their density, their average height and their diameter at breast height (dbh). Sites are selected in general along crono-sequences in order to obtain information regarding canopy age as a function of the HCI.
3. Frequency of cloud cover: since satellite and airborne techniques are affected by the frequency of cloud cover, sites are selected using statistical techniques based on meteorological data and cloud cover monitoring satellites. This allows for the selection of sites with little cloud cover during the sampling season.
4. Phenology: this is a key component of canopy studies on tropical dry forests. Sites are selected based on the percentage of tree species that have phenological stages (leaf-on to leaf off). This allows for the identification of spectral changes in canopy trees as a function of the time of year.

Political stability is a key issue in site selection as the personal well being of the researchers/students in the field has to be considered. The Santa Rosa National Park in Costa Rica and the Los Inocentes moist forests were eventually selected based on the four criteria mentioned above.

Equipment and installation

In general we use the following equipment in the field:
LI-COR 3100 and LI-COR 3000A (leaf scanners)
LI-COR 2000 (Leaf Area Index measurement, LAI-2000)

The LICOR-3100 is an area meter which provides a rapid, easy-to-use system for precise measurement of both large and small objects. It has dual capability of either 1 or 0.1 mm^2 resolution on the same instrument. The LI-3000A goes one step further and combines an easy to use microprocessor controlled readout console with the proven scanning technology of the LI-COR LI-3000 sensor head to provide a powerful system for portable nondestructive leaf area measurements. The LI-3000A also utilizes an electronic method of rectangular approximation to provide 1 mm^2 resolution. Leaf area, leaf length, average width, and maximum width are logged by the readout console as the scanning head is drawn over a leaf.

The LAI-2000 uses an innovative technique for making rapid, nondestructive measurements of leaf area index (LAI) and other plant canopy structure attributes such as Mean Tip Angle (MTA). Rapid sampling means that cost savings can be substantial when compared to direct measurements made with an area meter. Measurements can be made under a variety of sky conditions, and in canopies ranging from short grass to forest. These three instruments are supplied by Licor Inc. (www.licor.com). We also use high resolution Geographic Positional Systems (GPS) provided by Trimble (www.trimble.com), and two types of spectrometers provided by Analytical Spectral Devises (ASD, http://www.asdi.com/). These are the ASD-Pro and the ASD-Handled. ASD's spectroradiometers are ergonomic state-of-the-art instruments for quantitative measurement of radiant energy at all wavelengths from 350–2500 nm in radiance, irradiance, reflectance or transmission. Superior sensitivity combined with a 0.1 second scan time permits greater data collection in a shorter amount of time and faster movement between sites for researchers in the field. Airborne data is provided by several Canadian and US airbone companies. One thing to be aware of is that field spectrometers need cloud free days since they are used to measure the spectral reflectance of objects.

The choice of equipment is not based on cost, as LI-COR and ASD instruments are unique so there is not much choice for researchers. All equipment is mobile and thus does not require permanent installation. In general, we use portable towers as well, which take about 2 days to install.

Research projects

The diversity and richness of tropical environments has not been fully characterized using hyperspectral remote sensing techniques. Hyperspectral technology will allow us to move from the conventional 4 or 7-band (wide spectral resolution) approach to as many as 200+ bands (narrow spectral resolution). Hyperspectral remote sensing has been used over the last 10 years in boreal ecosystems for detailed ecosystem characterization, as well as to compose comprehensive forest resource assessments at the local scale. With the launching of new hyperspectral remote sensing satellites in the next 5 to 10 years there is set to be a large increase in the amount of data available. These satellites use selective tools and algorithms that are aimed at detecting canopies as a priority for tropical remote sensing research. The former is the key to the conservation of rare and endangered species such as the Great Green Macaw (*Ara ambigua*) – which depends on 14 tree species for its survival. In addition, the high spectral accuracy and resolution of hyperspectral data sets makes them ideal for the study of tree species spatial distribution. This knowledge can then be used to develop sampling schemes for bio-prospecting. Specific issues dealing with biochemical characteristics of tropical canopies can

also be studied using these datasets. The integration of information collected in the field using remote sensing techniques provides new insights into the nature and spatial distribution of canopies over large geographical areas.

I consider the use of hyperspectral tools a research priority for remote sensing of canopies. What is really needed in this field is the compilation of hyperspectral libraries of selective emergent tree species that could help not only with the calibration but also with the validation of future satellites. Currently, such standardized libraries of tropical trees and plants are not available. Comprehensive hyperspectral libraries of tropical tree species and plants will be critical to aid better detection and identification of the presence of specific trees that may have significant value for bioprospecting. They will also contribute to better monitoring of biological resources at larger scales without the necessity of performing extensive and expensive field campaigns.

I strongly believe that this is the future of remote sensing. It will be the basis for the identification of key species and their distribution in the landscape, and these techniques will be applied to conservation biology. The use of hyperspectral remote sensing techniques in the next decade will open up a new field in canopy research by means of the integration of airborne, satellite and field data (collected from towers, cranes or through destructive sampling of leaves). This new way of

a. Landsat Thermatic Mapper,
28.5 m resolution.

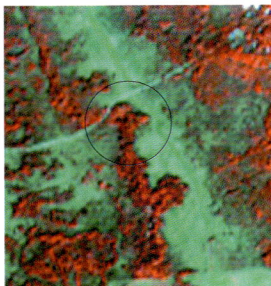

b. ASTER, 15 m resolution.

c. IKONOS, 4 m resolution

d. IKONOS, 1 m resolution

An eagle's view of a tropical dry forest canopy. Santa Rosa National Park forest island view through 3 different earth observation satellites and 3 spatial resolutions. The circle on the coarse resolution image lactes the 1-m data set.

dealing with canopy research issues not currently being addressed today will surface as a major force in the research field.

Costs
Costs are detailed as follows:

Table 1

COST		US$
Equipment	Licor 3000	6,000
	Licor 2000 (x2)	16,000
	GPS	50,000
	Spectrometers	200,000
Transportation		20,000
Staff	Technician	45,000 per year
	Graduate student support (x3)	50,000
Hyperspectral flights		60,000 per site (4 sites) = 240,000
TOTAL COSTS		627,000

Funding
Funding for our current work is provided by the Canada Foundation for Innovation (CFI), the National Geographic Society, and the provincial government of Alberta, Canada.

Safety
Snakes are the main problem when working in forest in the tropics, as long as they do not bite us we do not bite them!

There is always a need to have liability insurance, this is a pre-requisite of the University. Added to this non-University people are not allowed to participate in data collection and other sampling work in the field etc… due to insurance issues.

Contact details
Dr Arturo Sánchez-Azofeifa, Assistant Professor, Earth and Atmospheric Sciences Dept. University of Alberta, Edmonton, Alberta, Canada T6G 2E3.
Phone: + 1-780-492-1822, Fax: +1-780-492-2030
Email: arturo.sanchez@ualberta.ca, web: http://eosl.eas.ualberta.ca

Ecophysiology

By
Martin G Barker

Plant Ecophysiology and Nutrient Cycling

PLANTS dominate forest structure and function. In tropical forests in particular, plants often also represent a large proportion of the biodiversity (Allen 1996). Not surprisingly, therefore, many canopy researchers conduct studies involving aspects of plant form and function. In a recent survey of 112 canopy researchers, the two most common subject categories were plant ecophysiology and ecology (Barker & Pinard 2001). The prevalence of such studies reflects the central role of canopy plants in forest structure and function. Plant form and structure have a strong reciprocal relationship with the forest's physical environment. Plants influence and are influenced by microclimate, including ambient irradiance, air temperature and humidity, and carbon dioxide concentrations. These variables are highly dynamic in space and time (e.g. Fitzjarrald & Moore 1995). Microclimate gradients also have an important influence on the distribution of forest fauna and flora. Hence, there is a major incentive to quantifying plant-microclimate variables as part of our attempt to understand overall forest function.

Plants also play an integral role in forest nutrient cycles, which connect atmospheric and soil processes. Considerable attention has been directed at ground level processes such as litterfall, decomposition and groundwater leaching in forests. To fully understand nutrient cycling in forests, we need to study numerous canopy processes including leaching from leaves, atmospheric canopy deposition, nitrogen fixation by epiphytes, host-parasite nutrient relations and nutrient acquisition and storage by epiphytes (e.g. Nadkarni & Matelson 1991, Clark *et al* 1998). We also need to understand how nutrients circulate within trees (Barker & Becker 1995).

To understand the specific role of plants in forest structure and function, we need to understand how plants work (i.e. their physiology) in an ecological context (i.e. their ecophysiology). It has been said that ecophysiology is the study of physiology under the worst possible conditions. If this is so, then canopy ecophysiology probably is physiology measured under the worst, worst conditions.

In this chapter, I will develop the 'worst, worst conditions' theme by describing the constraints involved in conducting studies of plant ecophysiology and nutrient cycling in forest canopies. My comments are based on extensive personal experience of canopy studies in Borneo, Bolivia and the USA, and from numerous communications with other canopy plant researchers. Although I will emphasize problems, I will also identify solutions.

Plant ecophysiology in forest canopies: challenges and opportunities

Canopy plant ecophysiological research must be as rigorous as terrestrial plant research (Barker & Pinard 2001, Mitchell 2001). In other words, both the quantity and quality of data collected from the canopy should (ideally) not be diminished by the practical challenges of working in the canopy. I want to argue that this is neither unrealistic nor unfair. Why?

Firstly, canopy researchers need to use robust experimental designs in order to be able to publish results in high-quality journals. We cannot assume that reviewers and editors will make concessions simply because they recognize that canopy research is technically difficult. We have to accept that research will be evaluated in terms of objective scientific criteria, despite rather than because of difficulties of working in the canopy.

Secondly, many studies involve comparisons between canopy and understory measurements. To be comparable, different groups of data need to be sampled at an equivalent intensity. For example, additional effort may be needed to collect data from the upper crown, since it is difficult to reach. This may mean that some sample types will be under-represented.

I now consider a series of problems associated with investigating plant ecophysiology and nutrient dynamics in the canopy. These problems are presented roughly in the order in which the researcher might encounter them. However, as I will be arguing later, that all problems need to be anticipated and solved at the planning stage, before the fieldwork actually begins. We need to decide where to sample and when to sample.

Challenge 1: In situ or ex situ measurements?

How are we to reach parts of the canopy where we want to measure ecophysiological or nutrient processes? Most plant sampling in the canopy occurs, initially at least, in the outer crown area. This is where leaves, flowers and fine branches of trees, climbers and epiphytes are located. We need to decide whether to make measurements *in situ* or *ex situ*. *In situ* measurements are made in the canopy, using undisturbed material, while *ex situ* measurements elsewhere.

Ex situ measurements can be made on the ground or in the laboratory. Examples include foliar nutrient content and maximum assimilation rate (using a hydrated shoot). In these cases, shoots can be remotely brought down from low canopies using a pole pruner (Ingram & Lowman 1995). For higher canopies a projectile can be fired, for instance from a shotgun or catapult (slingshot). Alternatively, material can be collected by a single, rapid visit to the canopy (Mori 1995).

Another approach is to use one visit to the canopy to establish remote sampling positions. For example, pulleys can be placed in the canopy to allow sampling equipment (e.g. canopy litterfall traps) to be raised and lowered from the ground. Other sampling devices used in this way for plant ecophysiology could include cameras, for hemispherical or forest structure photography, or micrometeorology sensors (see Moffett & Lowman 1995).

We can also understand the ecophysiology of plants by reconstructing or simulating canopy conditions in the laboratory. This approach has been used for studies of branches (Schlesinger *et al* 1993, Parrish 1995) and small trees (Hilbert & Messier 1996). These methods simplify and control otherwise complex canopy environmental conditions.

In situ measurements are a greater challenge. We often need to measure plant ecophysiology and nutrient cycling under normal, ambient conditions. For example, net assimilation and stomatal conductance are both affected by local microclimate and the hydraulic conductivity of the plant. Such sampling must be done while the leaf is still attached to the plant. Similarly, nutrient gains and losses within the canopy can only be fully understood by small-scale canopy measurements. In these circumstances, the researcher must access the canopy. Discussion of the numerous problems associated with this occupies the remainder of the chapter.

Challenge 2: Canopy access

What access technique can we use to conduct plant ecophysiological or nutrient cycling measurements in the canopy? Numerous canopy access methods are now available (reviewed in Lowman *et al* 1993, Moffett & Lowman 1995, Lowman & Wittman 1996, Barker & Sutton 1997). The proliferation of methods has been both a cause and a consequence of advances in canopy science. Previously, canopy research had been seriously impeded by access difficulties.

The choice of canopy access method is influenced by several factors. These include cost, time (Zandt 1994) and forest type (Moffett & Lowman 1995). Decisions need to be made about the scale of sampling. For example, plant physiological measurements can be made at the leaf or shoot level, while canopy-atmosphere gas exchange measurements are made at the stand level (Cermák 1989). Different canopy access and sampling approaches are involved at each spatial scale.

Low-tech methods (Barker & Sutton 1997) have the advantage of being relatively inexpensive and mobile and can be used in remote forests without high-tech facilities. This can allow multiple, simultaneous sampling. For example, diurnal gas exchange measurements can be made on replicate trees, provided that enough instruments and personnel are available. Low-tech methods including the use of ropes and bole climbing equipment have been used in plant ecophysiological sampling in the outer canopy, though often with difficulty. Bole climbing methods (e.g. ladders, spikes, Swiss tree bicycle: see Barker and Sutton 1997) do not provide direct access to the outer crown. However, these methods can provide access to the lower canopy, from where rope methods can be used (see page section 1).

It is also often possible to bend branches to access a convenient sampling position, provided that this does not alter the experimental conditions. For destructive sampling, it is possible to extend the access range of SRT by using pole-pruners (Figure 1). I have used this technique to collect branches for xylem sap extraction, which can be done in the canopy (see Barker & Becker 1995).

With low-tech methods, it can be difficult to use delicate and/or bulky equipment. However, pulleys can be used to raise equipment (Perry 1978), and platforms can be used to establish a fixed, secure sampling position in the canopy (Nadkarni 1988). It often helps to have an assistant on the ground who can attach equipment on to a rope.

High-tech facilities, which include towers, walkways and cranes, are probably the preferred canopy access method by most ecophysiologists. Compared with low-tech methods, they offer greater sampling capacity, both in terms of the number and the duration of measurements that can be made (e.g. Zotz & Winter 1996). They are also a better base from which to use ecophysiological instrumentation. Combining canopy access methods (Barker & Pinard 2001) further extends the repertoire of techniques. Now that many access problems have been solved the primary concern of canopy researchers has shifted from access methods (Nadkarni & Parker 1994) to sampling issues (Barker & Pinard 2001, Bongers 2001).

Challenge 3: Replication and pseudoreplication

Reaching the canopy often provides only a partial solution to the problem of moving around in three-dimensional space, high above the forest floor. The next challenge is to reach specific sampling areas within the canopy. Access restriction can result in arbitrary or non-random and opportunistic sampling. This is a problem because replicates may be unrepresentative and insufficient (Zotz & Winter 1996, Barker & Pinard 2001). Since forest canopy environments are highly heterogeneous and complex spatial and temporal variability must be represented by replication (Lowman *et al* 1993). Pseudoreplication (Hurlbert 1984) is a serious risk in some canopy studies (Barker & Pinard 2001).

Adequate replication is highly subject-dependent (Bongers 2001). For plant ecophysiological studies, typical replicates are trees or plots. However, within tree sampling is valid if samples are independent. For example, individual branches may autonomous for water relations (Sprugel *et al* 1991), carbon economy or herbivory (Watson & Casper 1984). In these cases, leaves taken from different branches are functionally independent and can be regarded as replicates rather than pseudoreplicates (e.g. Barker & Booth 1996). However, the status of leaves as independent functioning units is debatable (Caldwell *et al* 1986; Leverenz 1988). In the understory, at least, leaves that are less than 2 m apart are probably affected by the same variations in irradiance (Becker & Smith 1990). Because of the complex environment within crowns, sampling areas need to be selected consistently and without bias, for example by using randomized branch sampling (Gregoire *et al* 1995, Zandt 1994).

For low-tech methods (e.g. ladders and ropes), the amount of time and effort required can seriously impede replication (e.g. Caron & Fleming 1995). Often in plant ecophysiology, we need to sample extensively, for instance to measure leaf water potential or gas exchange during a diurnal cycle. In such cases, immobile access techniques may only allow sampling from one tree per day. However, sampling can be increased by the use of

Figure 1 Collecting small branches for xylem sap nutrient analysis. Access in this case is by a combination of a canopy walkway, SRT and a pole pruner.
Photo: courtesy of Martin Barker.

extra personnel during periods of intensive fieldwork. Alternatively, sample days can be used as replicates, provided that experimental conditions (such as rainfall or irradiance) are consistent among days (later confirmed by analysis of covariance in the data).

Canopy access methods that restrict the researcher to a small area make replication difficult. Static structures such as towers and walkways allow the researcher to reach only a limited number of independent sample points (Nadkarni & Parker 1994). For towers, this problem can be overcome to some extent by using modular or movable structures and/or movable platforms (Lowman 1997, Bond *et al* 1999). The sampling range of walkways can be expanded horizontally (by building extensions) and vertically (e.g. by using SRT or a pole pruner). Similarly, ropes can be used to extend the vertical sampling zone of the canopy raft (Cosson 1995).

Even though cranes have a relatively large sampling capacity both in time and space (Zotz & Winter 1996). Many researchers, who use them report difficulties in achieving adequate replication (Barker & Pinard 2001). One problem is that, though a crane might provide access to say, 140 trees (Allen 1996), there may not be many individual (replicate) trees of any given species. However, if the focus is on functional groups, then different species could be counted as replicates.

Challenge 4: Disturbance and demonic intrusion

Plant ecophysiologists cannot ignore the risk of 'demonic intrusion'. This occurs when sampling methods, such as canopy access infrastructure, have an unintended effect on the measurements obtained (Barker & Pinard 2001). An example is the effect of large access structures on microclimate. For example, towers and walkways are necessarily located within gaps, which have a different microclimate than more intact areas of forest (Parker *et al* 1992). Another example is the heat radiation given off by the canopy raft (Koch *et al* 1994).

Disturbance effects are further compounded if they differ among access methods. For example, vertical patterns of irradiance within a forest may differ when measured from a tower or from the gondola of a crane (see Chazdon *et al* 1996). Thus the results obtained from studies using different access methods would not be comparable.

The best experimental practice is to recognize and rectify disturbance effects. For example, in one plant ecophysiology study (Bassow & Bazzaz 1997), a cherry picker was used for canopy access. To avoid engine exhaust interfering with gas exchange measurements, the motor was turned off during measurements.

Challenge 5: Logistics and hyperspace

Plant ecophysiological and nutrient dynamics measurements are highly dependent on time and place. In practice, this means that the time and location of canopy measurements usually needs to be defined, so that sufficient replication can be achieved. However, there are many logistical problems in sampling in three-dimensional canopy space (Nadkarni & Parker 1994). This may partly explain the lack of spatial replication that characterizes much canopy research (Stork & Best 1994). Since the canopy itself is structurally complex and heterogeneous, there are likely to be a correspondingly wide range of plant physiological responses. For example, canopy microclimate gradients are associated with variability in stomatal density and conductance, net assimilation, and nitrogen content in canopy leaves (e.g. Eliás

1979, Barker & Booth 1996). Nutrient content of xylem sap and leaves on the other hand is influenced by canopy height and aspect (Stark *et al* 1985).

Moving between sampling points in three dimensional canopy space adds a further time dimension, or hyperspace, to measurements (Kapos *et al* 1993). This could be a problem, especially for rapidly fluctuating variables because important temporal resolution can be lost from canopy measurements (Fitzjarrald & Moore 1995). Temporal sampling may also be constrained by high costs (Nadkarni & Parker 1994) and due to the sharing of facilities among researchers. In practice, real time sampling with adequate replication may be impractical in heterogeneous canopy conditions. However, many plant ecophysiologists scale up from a limited number of samples (e.g. of single leaves or branches), using models.

Challenge 6: Presentation and dissemination of results

Finally, we arrive back at our desks, exhausted but exhilarated from our fieldwork in the forest canopy. Now we have to analyze, interpret and present our data. If we have planned our sampling carefully and used a robust experimental design, there should be no difficulty in analyzing the data.

For some types of study, there is the additional challenge of representing values obtained from a three-dimensional sampling environment (Richards 1983, Popma *et al* 1988, Nadkarni & Parker 1994) e.g. the distribution of leaves in three-dimensional canopy space is hard to define but can now be modeled (Pearcy & Valladares 1999).

We should be explicit about our canopy access methods and about their experimental implications. We need to report our methods fully and develop common sampling protocols to allow comparisons among studies (Barker 1997). However, relatively few canopy research papers refer to sampling issues (Kapos *et al* 1993, Koch *et al* 1994, Baldocchi & Collineau 1994).

General guidance on field methods

Do

- Secure items that might be accidentally dropped from the canopy. This is to avoid inconvenience to the canopy researcher, damage to equipment, and danger to anyone below.
- Allow enough time to establish sampling positions, and the time required to move among them.

Don't

- Sacrifice safety for science, ever.
- Attempt to climb using low-tech methods without proper training. Every manoeuvre must be practiced first at ground level.
- Stand too close when someone else is using a line-firing device (e.g. slingshot, crossbow, rope launcher, etc).

Conclusions

We can transform challenges into opportunities by carefully planning our fieldwork in the canopy. We need to anticipate and recognize when experimental rigour may be compromised by difficulties in canopy access. Then we can think of creative solutions. Ignoring problems related to canopy access won't make them go away!

Because of the particular challenges of canopy research, we need to plan field-work to pre-empt problems. There is a need to anticipate the logistical and financial costs of fieldwork in the canopy. We may need to conduct cost/benefit analyses to decide what access method to use (Zandt 1994 Moffett & Lowman 1995). In studies for which consistent sampling effort among trees is important, we need to evaluate the implications of using different access methods on different trees.

Canopy research in plant ecophysiology and nutrient dynamics needs to be carefully planned to include effective sampling strategies. Research in forest canopies is a relatively young science. This means that are still experimental difficulties to overcome, but it also means that there are numerous research questions to answer.

Contact details

Dr Martin Barker, School of Forest Resources and Conservation, University of Florida, USA

Literature cited

Allen WH (1996). Traveling across the treetops. *BioScience* 46: 796–799

Baldocchi D & Collineau S (1994). The physical nature of solar radiation in heterogeneous canopies: spatial and temporal attributes. Pp. 21–71. In: Caldwell MM & Pearcy RW (eds). Exploitation of environmental heterogeneity by plants. Academic Press, San Diego, USA.

Barker MG (1997). An update on low-tech methods for forest canopy access and on sampling a forest canopy. *Selbyana* 18: 16–26

Barker MG & Becker P (1995) Sap flow and sap nutrient content of a tropical rain forest canopy species, Dryobalanops aromatica, in Brunei. *Selbyana* 16: 201–211

Barker MG & Booth WE (1996) Vertical profiles in a Brunei rain forest: II. Leaf characteristics of *Dryobalanops lanceolata* Burck. *J. Trop. For. Sci.* 9: 52–66

Barker MG & Sutton SL (1997). Low-tech techniques for forest canopy access. *Biotropica* 29: 243–247

Barker MG & Pinard MA (2001) Forest canopy research: sampling problems, and some solutions. *Plant Ecology* 153:23–38

Bassow SL & Bazzaz FA (1997). Intra- and inter-specific variation in canopy photosynthesis in a mixed deciduous forest. *Oecologia* 109: 507–515.

Becker P & Simth AP (1990). Spatial autocorrelation of solar-radiation in a tropical understory. *Agr. For. Meteorol.* 52: 373–379.

Bond BJ, Farnsworth BT, Coulombe RA & Winner WE (1999). Foliage physiology and biochemistry in response to light gradients in conifers with varying shade tolerance. *Oecologia* 120: 183–192.

Bongers F (2001). Methods to assess tropical forest canopy structure: an overview. *Plant Ecology* 153: 263–277.

Caldwell MM, Meister H-P, Tenhunen JD & Lange OL (1986). Canopy structure, light microclimate and leaf gas exchange of *Quercus coccifera* L. in a Portuguese macchia: measurements in different canopy layers and simulations with a canopy model. *Trees* 1: 25–41.

Caron GE & Fleming RA (1995). A simple method for estimating the number of seed cones on individual black spruce. *Can. J. For. Res.* 25: 398–406.

Cermák J (1989). Solar equivalent leaf area: an efficient biometrical parameter of individual leaves, trees and stands. *Tree Physiol.* 5: 269–289.

Chazdon RL, Pearcy RW, Lee DL & Fetcher N (1996). Photosynthetic responses of tropical plants to contrasting light environments. Pp. 5–55. In: Mulkey SS, Chazdon RL & Smith AP (eds.). *Tropical forest plant ecophysiology.* Chapman and Hall, New York.

Clark KL, Nadkarni NM & Gholz HL (1998). Growth, net production. litter decomposition, and net nitrogen accumulation by epiphytic bryophytes in a tropical montane forest. *Biotropica* 30: 12–23.

Cosson J-F (1995). Captures of *Myonycteris torquata* (Chiroptera: Pteropodidae) in forest canopy in South Cameroon. *Biotropica* 27: 395–396.

Eliás P (1979). Stomatal activity within crown of tall deciduous trees under forest conditions. *Biol. Plant.* 21: 266–274.

Fitzjarrald DR & Moore KE (1995). Physical mechanism of heat and mass exchange between forests and the atmosphere. Pp. 45–72. In: Lowman MD & Nadkarni NM (eds), *Forest canopies*. Academic Press, San Diego.

Gregoire TG, Valentine HT & Furnival GM (1995). Sampling methods to estimate foliage and other characteristics of individual trees. *Ecology* 76: 1181–1194.

Hilbert DW & Messier C (1996). Physical simulation of trees to study the effects of forest light environment, branch type and branch spacing on light interception and transmission. *Funct. Ecol.* 10: 777–783.

Hurlbert SH (1984). Pseudoreplication and the design of ecological experiments. *Ecol. Monogr.* 54: 187–211.

Kapos VG, Ganade E, Matsui & Victoria RL (1993). $\partial^{13}C$ as an indicator of edge effects in tropical rainforest reserves. *J. Ecol.* 81: 425–432.

Koch GW, Amthor JS & Goulden ML (1994). Diurnal patterns of leaf photosynthesis, conductance and water potential at the top of a lowland rain forest canopy in Cameroon: measurements from the *Radeau des Cimes*. *Tree Physiol.* 14: 347–360.

Leverenz JW (1988). The effects of illumination sequence, CO2 concentration, temperature and acclimation on the convexity of the photosynthetic light response curve. *Physiol. Plant.* 74: 332–341.

Lowman MD (1997). Herbivory in forests -from centimetres to megametres. Pp 135–149. In: Watt AD, Stork NE & Hunter MD (eds), *Forests and insects*. Chapman and Hall, London.

Lowman M, Moffett M & Rinker HB (1993). A new technique for taxonomic and ecological sampling in rain forest canopies. *Selbyana* 14: 75–79.

Lowman MD & Wittman PK (1996). Forest canopies: methods, hypotheses, and future directions. *Ann. Rev. Ecol. Syst.* 27: 55–81.

Mitchell A (2001) Introduction – Canopy science: time to shape up. *Plant Ecology* 153: 5–11.

Moffett MW & Lowman MD (1995). Canopy access techniques. In: Lowman MD & Nadkarni ME (Eds.) *Forest canopies*. Pp 3–26. Academic Press, San Diego.

Moreno J & Garcia-Martinez JL (1980). Extraction of tracheal sap from citrus and analysis of its nitrogenous compounds. *Physiol. Plant.* 50: 298–303.

Mori SA (1995). Exploring for plant diversity in the canopy of a French Guianan forest. *Selbyana* 16: 94–98.

Nadkarni NM (1988). Use of a portable platform for observations of tropical forest canopy animals. *Biotropica* 20: 350–351.

Nadkarni NM & Matelson TJ (1991). Fine litter dynamics within the tree canopy of a tropical cloud forest. *Ecology* 72: 2071–2082.

Nadkarni NM & Parker GG (1994). A profile of forest canopy science and scientists – who we are, what we want to know, and obstacles we face: results of an international survey. *Selbyana* 15: 38–50.

Parker G, Smith AP & Hogan KP (1992). Access to the upper forest canopy with a large tower crane. *BioScience* 42: 664–670.

Parrish JD (1995). Effects of needle architecture on warbler habitat selection in a coastal spruce forest. *Ecology* 76: 1813–1820.

Pearcy RW & Valladares F (1999). Resource acquisition by plants: the role of crown architecture. pp 45–66. In: Press MC, Scholes JD & Barker MG (eds). *Plant physiological ecology*. Blackwell Scientific, Oxford.

Perry DR (1978). A method of access into the crowns of emergent and canopy trees. *Biotropica* 10: 155–157.

Richards PW (1983). The three-dimensional structure of tropical rain forest. Pages 3–10 in Sutton SL, Whitmore TC & Chadwick AC (eds.) *Tropical rain forest: ecology and management.* Blackwell, London, UK.

Schlesinger WH, Knops JMH & Nash TH (1993). Arboreal sprint failure: Lizardfall in a California oak woodland. *Ecology* 74: 2465–2467.

Sprugel DG, Hinckley TM & Schaap W (1991). The theory and practice of branch autonomy. *Ann. Rev. Ecol. Syst.* 22: 309–334.

Stark N, Spitzner C & Essig D (1985). Xylem sap analysis for determining nutritional status of trees: Pinus menziessi. *Can. J. For. Res.* 15: 429–437.

Stork NE & Best V (1994). European Science Foundation -Results of a survey of European canopy research in the tropics. *Selbyana* 15: 51–62.

Popma J, Bongers F & Meave del Castillo J (1988). Patterns in the vertical structure of the tropical lowland rain forest of Los Tuxtlas, Mexico. *Vegetatio* 74: 81–91.

Zandt HS (1994). A comparison of three sampling techniques to estimate the population size of caterpillars in trees. *Oecologia* 97: 399–406.

Zotz G & Winter K (1996). Diel patterns of CO_2 exchange in rainforest canopy plants. Pp 89–113. In: Mulkey SS, Chazdon RL & Smith AP (eds), *Tropical forest plant ecophysiology.* Chapman and Hall, New York.

Forest Canopies and Climatic Change

By
Dieter Anhuf

THE forest canopy is where life meets the atmosphere and where a tremendous portion of the earth's biodiversity resides. The canopy is also the place where a enormous amount of sunlight energy is transformed, stored and transferred via litter-fall, herbivory and harvesting.

The significant role of tropical rainforests in regulating the regional surface energy and water balance and in maintaining the rich biodiversity of animal and plant species is increasingly endangered by human intervention such as deforestation and the conversion to agricultural land or pasture. There is an urgent need for information concerning the influence of terrestrial ecosystems and particularly of forests on biosphere-atmosphere interactions and their impact on the climate system and consequently their impact on climatic change.

More knowledge is needed to understand the climatological, ecological, biochemical and hydrological functioning of forests. Knowledge is also needed to assess the impact of land use change on these functions and the interactions between large forest blocks such as Amazonia, Congo, the boreal forests, and the earth system. Major aspects are the long term water vapour and carbon dioxide fluxes of different forest types around the world and in different climatic zones and their interactions with the climatic system.

The water and carbon cycle

The hydrological function comprises precipitation, evaporation, transpiration, surface and subsurface runoff, interception (precipitated water which is retained on the surface areas of vegetation), runoff from vegetation, and storage of water in soils, groundwater, and the oceans.

Atmospheric water vapour is produced by the evaporation of water from the oceans, from lakes, from land surfaces, and from the stomata of plants (transpiration). The energy required for this evaporation is provided by sunlight absorbed by these surfaces. Actual evapotranspiration accounts for 66 to 77 % of the net radiation equivalent (Szarzynski 2000). The water vapour is released to the upper layer of the atmosphere where it cools and condenses in clouds, forming water drops. The energy that was used for the evaporation of the water, is stored in the form of latent heat within the water vapour. Upon condensation, this latent heat is released. The water drops then return to the ground as precipitation.

Therefore the atmospheric water vapour above tropical rainforest regions is largely supplied by water evaporation from those forests. In dense tropical forests around 50 percent of precipitation is returned to the atmosphere as water vapour due to tree transpiration and interception. An equivalent of 20 to 25 percent of total evapotranspiration, evaporates from the surface of the plants (interception) (Anhuf *et al* 1999, Shuttleworth 1989). This means that 75 to 80 percent of total evaporation is achieved by transpiration from tree leaves. The remaining 50 percent of precipitation water either enters into the soil or flows directly into the ocean through rivers.

The large amount of evapotranspiration, i.e. the sum of transpiration and evaporation means that the level of rainfall is roughly the same throughout the entire Amazon and Congo Basins, even though the western (Amazon) and eastern (Congo) regions are several thousand kilometres away from the coast. A single water molecule can fall five to eight times onto the tropical forests of Amazonia before reaching the Andes via air currents coming from the Atlantic Ocean (Lettau et al 1979).

All forest blocks contain stores of carbon. Therefore, as mentioned above, forest conversion is a net source of carbon to the atmosphere, while recent measurements indicate that forest systems may be a net carbon sink. The importance of sequestration of carbon in growing forests and in old forest stands is unclear. These issues represent uncertainties in the global carbon balance and may influence the carbon dioxide concentration of the atmosphere and thus interact directly with the climate system.

Increasingly data concerning the hydrological and carbon cycles often forms the basis for political decisions on a regional and on a global scale within the framework of the discussions on climatic change. In this context, appropriate solutions can only be derived, if the complex ecological cycles of primary forests, their buffering ability, and sensitivity to disturbance are better understood. This improved knowledge can then be used to develop adequate strategies for secondary forest management.

Biotic and abiotic interactive processes

The complex interactions of the water and carbon cycles with climatological and ecological factors are of crucial importance for abiotic and biotic interactive processes. Within forest stands variations in microclimate produce a comprehensive spectrum of microclimatic niches which influences numerous biochemical, physiological, morphological and behavioural aspects of flora and fauna (Anhuf & Winkler 1999, Anhuf & Rollenbeck 2001, Endler 1993, Kira & Yoda 1989, Szarzynski 2000).

Climatological and hydrological parameters effect local ecological communities involving various stochastic and deterministic processes and form an essential part of the mechanisms maintaining terrestrial biodiversity (Linsenmair 1990). Since many ecological aspects of plants and animals are related to local abiotic conditions, measurements of meteorological and geophysical data provide comprehensive basic information on biodiversity. These parameters have to be examined in primary or near primary forest in order to evaluate changes in forest climate, water, carbon budget and ecology caused by conversion into secondary forests or agriculture.

To carry out such investigations canopy access has to be guaranteed (Wright et al 1994). There are different canopy access techniques, but for mid- and long-term observation and measurements the best tools are cranes, towers or platforms and walkways.

Cranes have already been discussed in detail earlier. Another access technique is the use of towers and platforms. These towers are exclusively erected to carry indispensable instrumentation which performs microclimatological and meteorological analysis and are not used for comparative ecological studies. There is still a tremendous imbalance between the facilities available for climatological, biogeochemical, and hydrological studies and those for ecological studies. This is not only due to lower costs for towers which range from 10,000 US$ for a small tower up to 20 m height, up to 45.000 US$ for larger ones up to 60 m height. But also because the expected results from climate change research are more spectacular because of their political, economical and societal impact than those from ecological studies.

But this opinion is changing, because recent data have shown that during the dry season transitional tropical forests are contributors of CO_2 to the atmosphere while they are a sink during the wet season. Comparable uncertainties also exist within evergreen rainforests. These differences are probably directly related to differences in forest structure and ecological function. Other questions are also arising which can only be answered if the different ecological systems within the tropical rainforest are better understood e.g. the role of biogenetic aerosols as cloud condensation nuclei, and a more general but crucial question: Is biodiversity important for climatic systems?

Instrumentation

To receive the information/data needed different approaches can be used. In general they can be divided into within forest and above canopy measurements. The instrumentation used depends largely on the funding available (specific details are obtainable from the "Large Scale Biosphere – Atmosphere Experiment in Amazonia (LBA)" (http://lba.cptec.inpe.br/). To record temperature and relative humidity probes have to be placed in the canopy at different heights depending on forest stratification. This can be done with the use of innovative rope and pulley systems. Additionally wind speed and PAR (Photosynthetic active radiation) are measured in the mid and upper canopy. Soil temperature is registered at up to five or eight different depths and several heat flux plates and soil moisture sensors were buried below the surface e.g. at 8 mm and 200 mm, respectively. Data is stored on data-loggers (e.g. Campbell Scientific, UK) operating with the same intervals as the above-canopy system.

The international standard for above canopy measurements are direct flux measurements, as described below:

Two micrometeorological configurations may be employed for simultaneous measurements within and above the forest. The above-canopy system comprises profile readings of air temperature, vapour pressure and wind velocity (recorded at different heights above the canopy, ideally following logarithmic distances as for example the 28.5, 31.5, 35 and 41 metre heights used at the Venezuelan Surumoni crane site (Figure 1). The same sorts of instruments can also be seen at a small tower experiment within a sugar cane plantation in the State of São Paulo, Brazil and this can be adapted for forest canopies. The instruments can be mounted on metal tubes 6m long, which are directly fixed to the crane tower. Dry and wet

Figure 1 Eddy Correlation tower in a woodland savannah in Southeast Brazil. Photo: Humberto da Rocha

bulb temperatures are measured with shielded and fully ventilated platinum-resistance psychrometers and wind speed with sensitive cup anemometers (both from Vector Instruments, UK). Data concerning radiative fluxes can be provided by two pyranometers (Kipp & Zonen, Netherlands) measuring the incoming and the outgoing short-wave radiation, a hemispherical pyrradiometer (Schenk, Austria) recording all-wavelength radiation and a quantum sensor (Sky Instruments, UK) recording PAR in the waveband 400–700 nm. Sensors are connected to a battery-powered data-logger (Campbell Scientific, UK) computing 10-min averages from 20-sec sampling intervals.

The study of the interactions between forests and atmosphere has recently made more routine by new developments in the eddy correlation technique. This technique, using an ultrasound anemometer provides new opportunities for directly estimating fluxes of carbon dioxide and water vapour at spatial scales of the order of kilometres. Eddy correlation is used for long term continuous measurements of mass and energy fluxes, to capture ecosystem seasonal dynamics, their response to extreme events (temperature, water stress etc.) and to provide a unique data–base of flux measurements to be used for model validation. The equipment and methodology should be harmonized by using common software and common instrumentation design in order to have a solid basis for site inter-comparisons (see: http://lba.cptec.inpe.br/).

In comparison to profile readings which often show gaps in data registration because of extreme atmospheric conditions the direct flux measurements produce more reliable and undisturbed data sets.

Table 1 contains the specifications for the instrumentation used to measure atmospheric variables and the sensor positions on a small tower experiment within a sugar cane plantation in the State of São Paulo, Brazil are listed.

Table 1 Specifications for sensors measuring variables

VARIABLE (DESCRIPTION)	UNIT	SENSOR / DEVICE	SENSOR POSITIONING
Air temperature	(°C)	Psychrometer CSI HMP 45 C	20 m
Air humidity	(g kg^{-1})	Psychrometer CSI HMP 45 C	20 m
Precipitation	(mm)	Texas 500	20 m
Wind direction	Degrees	RM Young	20 m
Wind speed	(m s^{-1})	RM Young	20 m
Global solar radiation	(W m^{-2})	LiCor 200x	incident, reflected 20 m
PAR radiation	(μmol m^{-2} s^{-1})	LiCor Quantum	incident, reflected 20 m
Net radiation	(W m^{-2})	REBS	20 m
Soil moisture	(m^3 m^{-3})	FDR CS615	05, 10, 40, 60,100, 150, 200, 250 cm depth
Soil temperature	(°C)	CSI T 108	05, 10 cm depth
Soil CO2 efflux	(gCO$_2$m^{-2} s^{-1})	IRGA EGM-2Ppsystems	Ground level
Latent heat flux	(W m^{-2})	Eddy correlation	20 m
Sensible heat flux	(W m^{-2})	Eddy correlation	20 m
Ground heat flux	(W m^{-2})	REBS HFT3	01 cm
Air total CO2 flux	(gCO$_2$m^{-2} s^{-1})	Eddy correlation	20 m

The needed equipment for a forest experiment can easily be summarized by multi-plying the specific sensors in dependence of the forest height and the number of the respective amounts of structural layers.

Suppliers

Campbell Scientific Ltd, Campbell Park, 80 Hathern Road, Shepshed,
Lougborough LE12 9GX, UK. Phone: 44.150960.1141, Fax: 44.150960.1091
Email: sales@campbellsci.co.uk; www.campbellsci.co.uk

Kipp & Zonen B.V., Röntgenweg 1, 2624 BD Delft, P.O. Box 507, 2600 AM Delft
The Netherlands. Phone: 31 15 269 8000, Fax: 31 15 262 0351
Email: kipp.holland@kippzonen.com, www.kippzonen.com,

Representative of Kipp & Zonen in Germany, Rolf Gengenbach – Messtechnik und
Datensysteme, Bachstrasse 9, 73269, Hochdorf, Germany.
Email: gengenbach@rg-messtechnik.de, www.rg-messtechnik.de

Delta-T Devices Ltd, 128 Low Road, Burwell, Cambridge, CB5 0EJ, UK.
Phone: +44 1638 742922, Fax: +44 1638 743155 www.delta-t.co.uk/
Sales and General Enquiries: sales@delta-t.co.uk
Technical Support: tech.support@delta-t.co.uk

Representative of Delta-T in Germany, UP – Umweltanalytische Produkte GmbH,
Bahnhofstr. 24, 03046 Cottbus, Germany.
Phone: +49(0)355/48554-0, Fax: +49(0)355/48554-15
Email: t.keller@upgmbh.com, www.upgmbh.com

LI-COR, inc., 4421 Superior Street, Lincoln, NE 68504, USA.
Phone: 402-467-3576, Fax: 402-467-2819
Email: envsales@licor.com, http://env.licor.com/

Representative of LI-COR in Germany, Heinz Walz GmbH, Eichenring 6,
D-91090 Effeltrich, Germany. Phone: +49-(0)9133/7765-0, Fax: +49-(0)9133/5395
Email: info@walz.com, www.walz.com/

Towers used around the world tend to be individualistic in design. It is advisable to contact local metal-processing industries to ask for a quotation and specific equipment. This proposal is made because of possible excessive transportation costs. Some example photos from towers in use in Brazil are included, which can be downloaded from: www.iag.usp.br/meteo/biota_fapesp/pro_cerrado_cana.htm or from http://lba.cptec.inpe.br.

Contact details

Professor Dieter Anhuf, Instituto de Estudos Avançados da Universidade de São Paulo, Travessa J, 374 05508-900 São Paulo, Brasil

Literature cited

Anhuf D & Winkler H (1999). Geographical and Ecological Settings of the Surumoni-Crane-

Project (Upper Orinoco, Estado Amazonas, Venezuela). *Anzeiger Math. Naturw. Kl. Abt. 1, Oesterreichische Akademie der Wissenschaften* 135: 3–25.

Anhuf D, Motzer T, Rollenbeck R, Schröder B & Szarzynski J (1999). Water budget of the Surumoni crane-site (Venezuela). *Selbyana* 20(1): 179–185.

Anhuf D & Rollenbeck R (2001). Canopy Structure of the Rio Surumoni Rain Forest (Venezuela) and its Influence on Microclimate. *Ecotropica* (in Press)

Endler JA (1993). The colour of light in forests and its implications. *Ecol. Mono.* 63: 1–27.

Gash JHC, Nobre CA, Roberts JM & Victoria RL (1996). *Amazonian Deforestation and Climate.* Wiley & Sons, Chichester, New York.

Kira T & Yoda K (1989). Vertical stratification in microclimate. In: Lieth N & Werger M (Eds.): *Tropical Rain Forests Ecosystems. Ecosystems of the World*, 14B, Elsevier, Amsterdam, 55–71.

Lettau HH, Lettau K & Molino LBC (1979). Amazonia's hydrologic cycle and the role of atmospheric recycling in assessing deforestation effects. *Monthly Weather Revue* 107: 207–237.

Linsenmair KE (1990). Tropische Biodiversität: Befunde und offene Probleme. In: *Verh.Dtsch.Zool.Ges.* 83, 245–261.

Nieder J, Engwald S, Klawun M & Barthlott W (2000). Spatial distribution of vascular epiphytes in a lowland Amazonian rain forest (Surumoni Crane Plot) in Southern Venezuela. *Biotropica* 32(3): 385–396.

Shuttleworth WJ (1989). Micrometeorology of temperate and tropical forest. *Phil. Trans. R. Soc. Lond.* B 324: 207–228.

Szarzynski J (2000). Bestandsklima und Energiehaushalt eines amazonischen Tieflandregenwaldes. *Mannheimer Geographische Arbeiten* 53, Mannheim.

Szarzynski J & Anhuf D 2001. Micrmeteorological conditions and canopy energy exchanges of a neotropical rain forest (Surumoni-Crane Project, Venezuela). *Plant Ecology* 153, 231–239.

Wright SJ & Colley M (1994). Accessing the Canopy: Assessment of biological diversity and microclimate of the tropical forest canopy. United Nations Environment Programme, Nairobi.

Maximising Data Collection

Training Parataxonomists to Survey Tropical Forest Canopies

By
Yves Basset

The parataxonomist trade

DANIEL Janzen and Winnie Hallwachs created the first parataxonomist course in 1989 in Costa Rica, in collaboration with the Instituto Nacional de Biodiversidad. The term 'parataxonomist' was coined as a parallel to 'paramedic', meaning that parataxonomists stand 'at the side' of taxonomists (Janzen *et al* 1993). In contrast to local informants, museum technicians and taxonomists (see discussion of each profession and its duties in Basset *et al* 2000), the expertise of parataxonomists is in collecting specimens, mounting them, preliminarily sorting them to morphospecies (i.e. unnamed species diagnosed with standard taxonomic techniques), and databasing the relevant information. Their work results in quality material that can be deposited in national collections and used for taxonomic studies. Although their role is more active than that of local informants (e.g. 'tree-spotters'), they cannot be seen as an alternative to professional taxonomists.

The term 'parataxonomist' has been used in different contexts and this is a source of confusion. Ultimately, all personnel involved in the collection and study of biological specimens may be viewed as 'parataxonomists': from local collectors, students, professional zoologists or botanists focusing on ecological studies, to taxonomists operating outside of their range of expertise. Here the emphasis is on local people living in relatively rural areas of the tropics and who have been specifically trained for parataxonomist duties by professional biologists, within the context of research projects.

The work of parataxonomists is usually most cost-effective when studying groups of small and species-rich organisms that may require microscopic observation and/or specific preparation for deposition in museums (e.g. herbaria specimens, pinned insects and preparation of their genitalia, etc.). Although parataxonomists often study plants (e.g. Beehler 1994), fungi (e.g. Bills & Polishook 1994), terrestrial arthropods (e.g. Janzen 1988; Longino & Colwell 1997; Novotny *et al* 2002), or benthic macroinvertebrates (Fore *et al* 2001), their skills may be useful for the study of many other taxa, including vertebrates.

Currently, despite much talking and the relative hype behind the term, sometimes even discernible in policy documents (e.g. UNEP 2001), one must acknowledge that few research projects routinely involve parataxonomists, especially in the tropics. Although the programme of work of the Global Taxonomy Initiative of the

Convention on Biological Diversity (see UNEP 2001) strongly encourages the development of parataxonomists, this advice has rarely been followed. A search in Biological Abstracts™ (1969–2001) provided only 6 records with the keyword 'parataxonomist'.

The question that must be asked is Why? Despite the appeal of the concept, many workers may still be suspicious of the quality of the data that may be recorded and archived by parataxonomists (see discussion of this in Fore *et al* 2001). By 'data quality' many workers imply 'data accuracy', but these are two different issues. Scientific methods in natural sciences differ from those in nuclear physics. For example, due to the high spatial and temporal heterogeneity of ecological factors in tropical rainforests, high numbers of replicates, even at the expense of lower accuracy, are likely to shed light on interesting biological patterns. Although parataxonomist work may result in lower accuracy of data, data quality may indeed be higher than the traditional work of the lone scientist(s), due to increased replicates and additional side-experiments.

Advantages and rewards of the parataxonomist strategy

These advantages have been discussed in detail elsewhere (Novotny *et al* 1997; Basset *et al* 2000) and can be summarized as follows:

1. Efficiency of fieldwork is comparable to that of professional biologists and allows collecting at simultaneous locations with a higher number of replicates. The amount of biological material collected may be considerable (e.g. Novotny *et al* 2002) and sampling efficiency is significantly higher in projects working with parataxonomists than in those not relying on them (Basset *et al* 2000). The feasibility of more ambitious projects with complex protocols is enhanced and allows, for example, the implementation of simultaneous inventories and biological monitoring within the study areas.

2. Preparation of high-quality biological material ready for deposition in permanent systematic collections may also be comparable to that of museum technicians. Local preparation of specimens may sometimes be advantageous. For example, reared moths and butterflies killed by freezing just prior to mounting often represent better specimens than those collected by e.g. light trapping.

3. The ecological information associated with the biological material may also be considerable. Knowledge of the environment by local people may be essential and profitably integrated in research projects. In addition, parataxonomists can be trained to perform side experiments that may be of high benefit for the interpretation of distribution data (see Novotny *et al* 1999 for such an example).

4. The time-lag between the initiation of the study and the publication of results, is often rather long for studies of megadiverse systems (e.g. Erwin 1995), and may be significantly reduced (see Basset *et al* 2000). This may be a particular advantage for conservation studies in which time is pressing and the need for action high.

5. The indirect but positive effects of local involvement in research projects should not be underestimated. Involvement of village communities in ecological research may demonstrate to them the value of undisturbed forests on their lands. Collateral education of local people by fellow parataxonomists may also be significant.

How to improve the training and accuracy of parataxonomists

The correlation between the data generated in sorting insect material to morpho-species by non-specialists (parataxonomists) and similar data obtained in sorting to species by expert taxonomists depends crucially on the standards of training and support, including provision of identification aids and quality control (e.g. Cranston & Hillman 1992; Fore *et al* 2001). Several tactics can ensure successful training of parataxonomists. First, the feedback of professional taxonomists during the life-time of the fieldwork is essential, in order to validate the morphospecies assignment of problematic groups (but not necessarily to name or describe species at that time). Second, recent developments in computer hardware make digital photography a useful and relatively cheap tool. Digital pictures of specimens and characters can be routinely included in sophisticated databases, and this information can be circulated readily among colleagues over the internet. Large public databases, such as Ecoport (www.ecoport.org) and taxonomic tools are also beginning to be widely available on the internet. All of these modern tools can greatly enhance the ability of parataxonomists to work efficiently and accurately.

Parataxonomy and biological monitoring

Biological monitoring usually implies specific protocols, such as nested or replicated samples, time-series or Before/After-Control/Impact designs (BACI). Long-term monitoring is best achieved with non-destructive, non-disturbing methods producing seasonal and annual replicates of the same sampling units. These protocols call for prolonged stays in the field, and parataxonomist input. For example, with the help of parataxonomists, we were able to achieve in Guiana one of the first BACI experiments proving unequivocally the influence of selective logging on rainforest insects (Basset *et al* 2001).

With the help of parataxonomists, some of the most time-consuming but inexpensive sampling methods become viable alternatives to more expensive methods of biological monitoring. It also becomes feasible to include several taxa or guilds within the sampling protocol. This represents a much more promising strategy than using the services of experts or students to monitor a species-poor taxon over relatively short periods, a bygone era of tropical bioinventories (Takeuchi & Goldman 2001).

Conclusion: parataxonomy and canopy research

The training and work of parataxonomists could be profitably put to use in conservation biology, especially in biological monitoring, and should be more often considered when planning such projects. This strategy may be particularly effective with invertebrate taxa, but not limited to them. To date, the author knows of no project specifically targeting tropical forest canopies and routinely including the work of parataxonomists. This is most unfortunate. Parataxonomists can be easily trained to use single rope techniques to access the canopy (Basset *et al* 2000) or they can use other facilities for canopy access. Parataxonomists could also represent key elements for efficient programme's of invertebrate mass-sampling, such as canopy fogging and light-trapping. Networking within various countries, groups of local parataxonomists monitoring functionally diverse canopy taxa, would also appear to be the way forward for efficient survey and biological monitoring of tropical forest canopies.

Contact details

Dr Yves Basset, Smithsonian Tropical Research Institute, Apartado 2072, Balboa, Ancon, Panama City, Republic of Panama. Email: bassety@tivoli.si.edu, Fax +507 212 8148.

Literature cited

Basset Y, Charles EL, Hammond DS & Brown VK (2001). Short-term effects of canopy openess on insect herbivores in a rain forest in Guyana. *Journal of Applied Ecology*, 38: 1045–1058.

Basset Y, Novotny V, Miller SE & Pyle R (2000). Quantifying biodiversity: Experience with parataxonomists and digital photography in New Guinea and Guyana. *BioScience*, 50: 899–908.

Beehler BM (1994). Using village naturalists for treeplot biodiversity studies. *Tropical Biodiversity*, 2: 333–338.

Bills GF & Polishook JD (1994). Microfungi from decaying leaves of *Heliconia mariae* (Heliconiaceae). *Brenesia*, 41–42: 27–43.

Cranston P & Hillman T (1992). Rapid assessment of biodiversity using 'biological diversity technicians'. *Australian Biologist*, 5: 144–154.

Erwin TL (1995). Measuring arthropod biodiversity in the tropical forest canopy. *Forest Canopies* (eds Lowman MD & Nadkarni NM), pp. 109–127. Academic Press, San Diego.

Fore LS, Paulsen K & O'Laughlin K (2001). Assessing the performance of volunteers in monitoring streams. *Freshwater Biology*, 46: 109–123.

Janzen DH (1988). Ecological characterization of a Costa Rican dry forest caterpillar fauna. *Biotropica*, 20: 120–135.

Janzen DH, Hallwachs W, Jimenez J & Gamez R (1993) The role of parataxonomists, inventory managers, and taxonomists in Costa Rica's national biodiversity inventory. In *Biodiversity Prospecting: Using Generic Resources for Sustainable Development* (eds Reid WV, Laird SA, Meyer CA, Gamez R, Sittenfeld A, Janzen DH, Gollin MA & Juma C), pp. 223–254. World Resources Institute, Washington.

Longino JT & Colwell RC (1997). Biodiversity assessment using structured inventory: capturing the ant fauna of a tropical rain forest. *Ecological Applications*, 7: 1263–1277.

Novotny V, Basset Y, Auga J, Boen W, Dal C, Drozd P, Kasbal M, Isua B, Kutil R, Manumbor M & Molem K (1999). Predation risk for herbivorous insects on tropical vegetation: a search for enemy-free space and time. *Australian Journal of Ecology*, 24: 477–483.

Novotny V, Basset Y, Miller SE, Allison A, Samuelson GA & Orsak LJ (1997). The diversity of tropical insect herbivores: an approach to collaborative international research in Papua New Guinea. In *Proceedings of the International Conference on Taxonomy and Biodiversity Conservation in the East Asia* (eds Lee BH, Choe JC & Han HY), pp. 112–125. Korean Institute for Biodiversity Research of Chonbuk National University, Chonju.

Novotny V, Basset Y, Miller SE, Drozd P & Cizek L (2002). Host specialisation of leaf-chewing insects in a New Guinean rain forest. *Journal of Animal Ecology*, in press.

Takeuchi W & Golman M (2001). Floristic documentation imperatives: some conclusions from contemporary surveys in Papua New Guinea. *Sida*, 19: 445–468.

UNEP (2001). *Convention on Biological Diversity. Subsidiary Body on Scientific, Technical and Technological Advice.* Sixth meeting. Montreal, 12–16 March 2001. Recommendation VI/6. The Global Taxonomy Initiative: programme of work. Published at http://www.biodiv.org/recommendations/default.asp?lg=0&m=sbstta-06&r=06.

Volunteers: Their Use and Management

By
Roger L. Kitching
&
David C Shaw

MANY of the tasks which confront the canopy biologist demand activities that go beyond those which can be provided by the Principal Investigator, professional assistants and graduate students. This factor taken together with the fact that many of the field and laboratory tasks require diligence and application but NOT professional-level expertise point to the need, and indeed, the desirability of using volunteers. This is an inexpensive tool which can increase productivity many-fold. The key words are 'properly managed' – without careful forethought volunteers may be a millstone around the neck of the scientist, and at worst can ruin a piece of research such that it must be repeated or even abandoned. This is a rare event and, in our experience, is more likely the result of poor leadership and management rather any basic inability on the part of the volunteers.

There are research activities for which volunteers are clearly not appropriate – where individual creativity and responsibility remain the vital ingredients. These solo efforts are however rare in field ecology – creativity and responsibility are still required *plus* the people skills which turn a crowd into a team.

Do you need volunteers?

If you think the answer is 'yes', ask yourself three key questions.

How many people do I need? Having too many volunteers is certainly worse than too few. Inadequate supervision may lead to unusable results from *all* the volunteers. With too few, the results that are generated are still likely to be of sufficient quality to be useful.

How much will they cost? There is a monetary cost to volunteers. They must be transported, insured, equipped, housed and fed. So, although you will not be paying them a salary they will cost you money – and for a team of ten or twenty, for example, this can add up.

What am I prepared to give them? Volunteers do not help in a research project purely out of a sense of philanthropy: they expect something in return. Inevitably the currency in which they seek payment is knowledge and inspiration. Volunteers have a genuine interest in what they are doing: they may be bored students, retirees seeking

Figure 1 Earthwatch Institute screens and deploys over 3,500 volunteers to peer-reviewed field projects each year. Their efforts generate funds for science and accelerate data collection. Photo: courtesy of Earthwatch

to pick up on previous interests, or people seeking a change from their regular careers. Time must be budgetted in which you share with them your excitement in the project, knowledge of the background science, even your scientific views in other areas.

Finding your volunteers

For anyone considering the use of volunteers for research they should remember the saying, "There is no such thing as a free lunch!"

Volunteers can be incredibly productive and enhance one's research program. But they require significant forethought and planning to be utilized in an efficient and productive way. The type of work that volunteers can accomplish must also be thought through carefully, along with the type of training that will be provided. The cost of using volunteers may seem low and this can be a large influence on deciding to go down the volunteer route. However it must be remembered that if you are spending all your time assisting volunteers, you may find no time for the work you thought you were going to get done. These issues will be looked at in more detail later in this section.

The type of model for volunteer use that works best, is that which screens potential volunteers before accepting them. Various internship programs work along this method and can produce exceptional volunteers. The United States Forest Service has a working program where volunteers are paid a daily stipend for food, but the volunteers must apply for the position like any other job. This works well, because the researcher can choose the person who appears most reliable and exhibits at least some aptitude for the project.

In the United States there is also a programme called, "Americorps", modelled along the lines of the Peace Corps. An individual applies to join Americorps and commits to one year of service for basic room and board and a college or vocational training stipend. Americorps volunteers can join field teams or work on individual placements. An individual placement located at the Wind River Canopy Crane Research Facility for 9 months followed an interview of the person suggested by the Americorps leaders prior to accepting the position. The experience was very productive. Teams have also been used at the Crane Facility for such tasks as trail maintenance, weed control, and litter sorting; all on a short term basis.

Perhaps the most successful use of volunteers has been accomplished by the Earthwatch Institute (www.earthwatch.org) and the researchers who cooperate with them (Figure 1 and 2). This organization is in the forefront of the constructive use of adult volunteers. Researchers apply to Earthwatch for grants to support a research team for a project of specified length. Project applications are then reviewed by a committee and, if approved, receive support in the form of volunteers which the organization recruits and funds. This approach has many advantages in removing from you, the researcher, the need to do your own recruiting. The volunteers pay for their own transportation and the costs of their support. The contribution they make also goes towards funding the costs of the research programme. Because the Earthwatch Institute requires a well thought through, peer reviewed, research design, and programme of how the volunteers will be used before any support is granted, it makes for a good model. Usually the projects must include some interesting element, such as an exotic location, but more directly a conservation biology application that will give volunteers a feeling of contributing in a positive way to the conservation of life on earth.

A variation on the usual Earthwatch 'expedition' uses high school students (under 18 years of age) in a program called the Student Challenge Awards. This program requires high school students who excel in humanities, but

Figure 2 Volunteers learning about vegetation during a bushwalk on an Earthwatch Expedition in Lamington National Park, Australia
Photo: R. Kitching

have an interest in science, from throughout the USA to submit applications to an Earthwatch director. The Institute then chooses the best applicants and assigns them to one of a number of projects. The students are keen to learn and work very hard. Problems can of course emerge, especially when sex becomes an issue, but this is something that just needs to be dealt with up front. Keeping the sexes in separate living quarters is helpful, but not always possible. Earthwatch does provides funds for Graduate Student Mentors to live with the students. However, even these mentors can't always stay awake!

Volunteers are also readily recruited, by academics at least, from their undergraduate classes. More highly motivated students rapidly realise that success in their chosen career will depend not only on their grades but on the other activities, such as volunteer work, that they can put on their embryonic *curricula vitae*. Again there are some pitfalls especially with the more junior students. They often promise more in time than they can deliver. Further, the routine aspects of field or laboratory work may be simply too much for those used to instant gratification: pinning and labelling twenty insects may be interesting, but 2000 (or the prospect thereof) may simply send them rushing back to the student bar, or their favourite computer game. For them this may be a lesson well learned as they appreciate the realities of scientific research: for the investigator it can be a complete pain as it is very difficult to keep track of lots of partly sorted or processed samples or mini-projects. In the Kitching laboratory we start student volunteers on trial work (on spare samples, less crucial activities) before we transfer them to 'main-line' work. It is an advantage to begin such activities with a bout of residential field work, where alternative activities are few.

Volunteers can also be recruited from citizen groups, non-Governmental organisations and the general public. When running a field expedition in a particular location interested locals sometimes turn up and some are more than willing to help. This can be an enormous advantage when working in a region where local knowledge and acceptance is important. Never turn away or fob off a passing visitor: they may have hidden skills. For specialised work, volunteer groups may already exist (e.g. 'Friends of the Museum/Herbarium', 'Conservation Volunteers', 'Honorary Rangers' etc) and these can be used with ease.

Negative experiences with volunteers can occur, especially when volunteers are accepted without prior interviewing or screening. People may have an agenda for volunteering that you are not aware of, and this can be quite problematic if it comes as a surprise. In addition, volunteers that don't realize the type or amount of work involved in their volunteer project may get upset or walk off the job, leaving you with no recourse. If you have spent a lot of time training, it can turn out to be a significant waste of your limited time.

Having said that, the use of volunteers can be a very rewarding experience and can massively enhance the research that can be achieved on a project. They are a useful tool for many researchers but, like all tools, they must be used properly.

Tasks and skill levels

Volunteers will come to you with a great variety of skill levels. Sometimes of course the volunteers arrive with skills in the actual area of research – their college training may have been in ecology, or they have worked with other scientists. More often than not though you will need to train your volunteers. Even the 'specialists' will need to be trained in your particular protocols. Such training needs to be done with great sensitivity. These are volunteers, remember, and if they don't like your style or demands they can simply leave – or, worse, continue to eat your food, occupy the spaces you are paying for, but simply do no useful work. With most volunteers the speed of skill acquisition is high. Volunteers are with you because they *want* to know about plants or insects or birds or whatever. Over a number of expeditions, one of us has successfully trained within a couple of days, 98% of 200 Earthwatch volunteers, to sort mixed arthropod samples to Order level. Of course continued supervision was needed to check difficult choices, and some performed better than others, but all produced usable data and ended with a real sense of a new skill acquired – as well as gaining access to what was, for them, a new world.

Volunteers also come with skills in other areas, and this can be turned to great advantage. In Australia we used a volunteer who was a civil engineer, to direct a topographic survey of the research plot being studied. Until appraised of his particular skills there had been no intention to carry out such a survey: the opportunity presented itself and the result has proved invaluable. Others have no technical skills but have vital people skills. They will often be the ones to appraise you, the team leader, that 'x isn't happy', or 'y is having a hard time coping with particular activities (or other members of the team, or even with you the leader)'. Such information must be dealt with and the 'people people' are often the ones who can help you do it.

Volunteers will be thirsty for other knowledge. In residential trips we usually have a daily pre-dinner lecture (or other group activity). Topics will be a matter for personal choice BUT remember that broad-brush overviews with lots of anecdotes are likely to go down much better than the detail that belongs in a seminar. High-tech visual aids are unlikely to be available in the field so plan to illustrate your talk using white or black boards (remember how?).

Safety and related matters

In the field you, the principal investigator, become responsible for the safety and well being of your volunteers. Some steps can be taken beforehand to minimise any risks involved. Seek and pass on advice on vaccinations or prophylactics necessary for your location. Inform your volunteers in writing about their medical needs in this regard. Ensure your institution (or someone) has insurance policies which protect you and your volunteers in the case of accident. Carry a major first-aid kit and ensure one of your staff is trained in its use. Check if your volunteers are on medication, what they are allergic to and whether they have serious phobias which will interfere with the work or their safety. You also need to know the civil procedures for seeking assistance in the case of a major emergency.

The first field task should be a safety briefing – about the forest and its dangers (remember: broken ankles and thorn-torn flesh are much more common than snakebite). This can also be a good time to encourage your team to have minimum impact on the field site as well as vice-versa. Establish rules about footwear, swimming, solo excursions (not generally a good idea) and what to do if there is an accident.

Extreme caution is called for when safety is an issue. Using volunteers for tree climbing or on other access systems should be well planned and thought through before hand. Training is a very critical component of this type of volunteer use. Perhaps it is best to use volunteers as support personnel, for processing of collected materials or as ground support for the climbers. At the Wind River Site volunteers have been used in support of forest canopy research. They have assisted in creating permanent tree/stem plots (12 ha plot), measuring and creating forest crown projection and canopy profiles, assisting with understory vegetation sampling, surveying canopy attributes of trees, surveying mistletoe in tree crowns, taking tree measurements such as height, diameter, crown size, and digging soil pits. In addition, these volunteers have used very sophisticated equipment, including Nikon total survey stations, GPS equipment, survey lasers, compasses, and computers. It is important to realise that accidents do happen but your task is to ensure that they are not made more likely or more serious by anything you, the leader, did or did not do.

Day-to-day management

A key event for the day on any field activity is the morning briefing. It was here, in Earthwatch expeditions in Australia and elsewhere that we reviewed exactly what had been achieved in a field trip to that time – what had gone wrong the previous day (there's always something), where we should be on the overall schedule, modifications to previous instructions, and so forth. Tasks for the day are assigned – after a few days it will become clear who is 'good' at a task and can take a leadership role. Yet everybody should get to try all tasks and to take a lead in at least some of them.

For complex survey work (often the sort which demands lots of volunteers) labelling protocols are the beginning, continuation and end of successful work. An unlabelled, or wrongly labelled sample takes hours to figure out and sometimes is better just discarded and re-collected. Simple code systems explained to your team every day, is usually successful but do not assumed that volunteers will have the 'feel; for commonplace abbreviations with which you may have become (over-) familiar.

Morning briefings are also an opportunity to draw the team's attention to interesting interim findings, natural historical sightings, even humorous incidents.

Dos and Don'ts

DO

- talk frequently to your volunteers about the work, their roles, their responses and their state of mind
- praise achievements however minor, as they occur: milestones may be marked by chocolate, wine or other 'sweeteners'
- do your homework on your volunteers' backgrounds – their training, previous experience, hobbies etc
- keep an eye (and ear) open for interpersonal differences among volunteers or others
- check constantly for correct labelling, sorting, storage and so on
- prepare written descriptions, picture keys, and any other helpful visual aids
- tell your team why what they are doing is important (if you don't know don't proceed with the project)
- admit you're wrong (when appropriate) or that someone has just shown you a better way of doing something, or that you have decided to abandon part of the work as impossible or inefficient, or just too time-consuming: volunteers are very understanding people as long as they know why change is occurring
- ensure members of your staff are aware of the principles of volunteer management.

DO NOT

- criticize one volunteer in front of another
- always favour the more competent over the less competent
- expect tasks of volunteers in which you do not yourself participate
- assume prior technical knowledge – of how a microscope works, or the difference between an insect and a mite, or whatever
- expect effort beyond years or physique

Contact details

Professor Roger Kitching, Australian School of Environmental Studies, Griffith University, Brisbane, Qld 411, Australia. Email: r.kitching@mailbox.gu.edu.au

Dr David Shaw, College of Forest Resources, University of Washington, 1262 Hemlock Road, Carson, Washington 98610 USA. Phone 509-427-7028. Fax 509-427-7037. Email: dshaw@u.washington.edu.

11

Data Management

Lasers in the Jungle: The Forest Canopy Database Project

By
alini M Nadkarni
&
Judy B Cushing

T HE forest canopy is of critical importance for a variety of life processes in our biosphere. Defined as "the combination of all leaves, twigs, and small branches in a stand of vegetation, including the air and interstices of the foliage", elements of the forest canopy house the photosynthetic machinery of the forest, influence the exchange of energy and matter with the atmosphere, control the microclimate at various scales, and maintain habitat for wildlife (Parker 1995). Forest canopy studies bear directly upon three of the most pressing environmental issues of the new millennium: the maintenance of biodiversity, the stability of world climate, and the sustainability of forest.

Both the types and amounts of canopy data are changing rapidly. In the past, scientists working alone with simple rope-climbing techniques generated studies that produced fairly small data sets. However, the recent access innovations permit multiple teams of scientists to work within the same volume space of the canopy. Canopy scientists now have to deal with more and new kinds of data, and the need to share it. Data collected by canopy research teams will be useful to other scientists (e.g. geographers and land use managers), just as data emanating from allied fields could aid forest canopy researchers.

The project

Historically, canopy scientists have been notorious for independent ways of taking, storing, and analyzing data. In 1993, our team of forest canopy ecologists and computer scientists received a planning grant from the National Science Foundation's (NSF) Database Activities Program. The project was to bring together forest canopy researchers, quantitative scientists, and computer scientists to work towards establishing methods to collect, store, display, analyze, and interpret three-dimensional (3-D) spatial data relating to tree crowns and forest canopies. We created a self-sustaining non-profit organization, the International Canopy Network (ICAN) to assure that network activity would continue beyond the life of the NSF grant (Nadkarni *et al* 1995). There are regular regional, national, and international meetings, workshops and symposia on canopy topics.

We also conducted a survey of over 350 canopy researchers and evaluated potentially applicable information models and software tools used in allied fields (Nadkarni & Parker 1994). This led to the organisation of a multidisciplinary workshop for canopy scientists and database/computer scientists. These activities:

1. Identified important questions under study in the emerging field of canopy research;
2. Formulated a number of key forest structure-function relationships that are currently poorly understood due to lack of database tools, and
3. Generated common ground for joint research by canopy researchers and database scientists

The rationale

The conclusion of both the survey and the workshop was that understanding forest canopy biota and processes was not limited by canopy access (as we had anticipated), but rather by two characteristics of canopy data:

1. Lack of quantitative tools that allow canopy researchers to analyze the complex three-dimensional spatial data associated with forest canopy studies, and
2. Lack of harmonized data sets – forest canopy researchers have tended to collect data in non-comparable formats

Although canopy datasets are increasing in number and variety, they tend to be anecdotal and descriptive and are not readily combined with others to expose general patterns or rules. Much of this is because canopy researchers have been preoccupied with detailed descriptions of particular environments, stands, processes, and study objectives. At the brink of the new millennium then, the study of the canopy of forest ecosystems is being held back by the lack of data management tools. The relative youth of the field with its lack of entrenched methods, legacy datasets, and conflicting camps of competing groups provides a unique opportunity for integrating data management and analysis tools into the research process. The sociology of the discipline is conducive to sharing data; researchers appear openly communicative and supportive of each others' work. Thus, forest canopy studies serve as an excellent arena to generate database tools that could also serve other fields of ecology and science.

In this commentary, we describe one fundamental part of the development of the field canopy studies, the Forest Canopy Database Project. The project was to develop a database and database tools to enhance the ability of researchers in one emerging and interdisciplinary field – forest canopy studies – to collect, analyze, link, and archive data. This capability will speed the development of the field to more efficiently address both intellectually stimulating and environmentally pressing questions of interest to academics, policy-makers, and the general public. We anticipate that the database and tools can serve as an exemplar for other interdisciplinary and emerging fields of science.

The databases and tools

To date, the computer database has taken two pathways. The first piece is our web-based centralized 'Big Canopy Database'. This database holds information, field data, and images of use to canopy researchers, educators, and conservationists, including lists of researcher contacts, research projects, study area descriptions, images, canopy-dwelling taxa, visualization and analysis programs, meetings, training programs, equipment and safety descriptions, and scientific and popular citations. A prototype is available for viewing at: www.evergreen.edu/canopydb

The second piece is a web-based program called 'DataBank', that will allow canopy researchers to search for and download field data submitted by other researchers, design field databases and download them for their own use, and to document and archive their own databases. The system thus builds new databases from database components that 'fit' canopy data. We term these components 'templates'.

DataBank currently contains datasets from six different canopy projects. To submit data to the database, a researcher from each study works directly with a database technician to provide metadata and to structure his/her data to fit one or more existing field data templates, or to generate a new template for novel data types. We anticipate that after a number of studies are entered, a finite number of data templates will be available, and researchers joining the database will find what they need within the program, obviating the need for an intermediary. The current DataBank prototype is implemented in SQL Server, Microsoft's Active Server Pages (ASP) and HTML. We are currently enhancing that prototype, using SQL Server but with Java rather than ASP.

Our efforts to create a database for the canopy research community will help push forward this emerging field of science. We also believe that our efforts could be viewed as a model for other emerging areas of ecology where data-linking and data-sharing can be effective in integrating results from different studies. We are seeking input from researchers in the field of canopy studies to contribute to the database, and from those outside the field who may have insights into making this process more efficient and productive.

Contact details

Author for correspondence*:
Nalini Nadkarni*, Tel: (360) 867-6621, Fax: (360) 866-6794, Email: nadkarnn@evergreen.edu.
Nalini M. Nadkarni & Judy B. Cushing, The Evergreen State College, Olympia, Washington 98505 USA

Literature cited

Nadkarni NM, Parker GG, Ford ED, Cushing JB & Stallman C (1996). The International Canopy Network: A pathway for interdisciplinary exchange of scientific information on forest canopies. *Northwest Science* 70:104–108.

Nadkarni NM & Parker GG (1994). A profile of forest canopy science and scientists – who we are, what we want to know, and obstacles we face: results of an international survey. *Selbyana* 15:38–50.

Parker GG (1995). Structure and microclimate of forest canopies. Pp. 73–106 In: *Forest canopies*, Lowman MD & Nadkarni N (eds). Academic Press, San Diego, USA.

Parker GG, Smith AP & Hogan KP (1992). Access to the upper forest canopy with a large tower crane. *BioScience* 42:664–670

The Way Forward – The Canopy Grid

By
Robert
Muetzelfeldt
and Judy Cushing

THE Global Canopy Programme (GCP) (see section 12) aims to integrate forest studies across the world into a ten year linked programme of research, conservation and education, focussed on understanding the critical role of forest canopies in biodiversity and climate change. It also aims to identify the societal benefits of forest canopies, and to transmit information to key stakeholders.

In order to meet these aims, the GCP has identified Data Management, Modelling and Communication as a Core Project Area. This is based on recognition of the importance of data sharing between canopy scientists, the integration of data and understanding in simulation models, and the communication of information within the forest canopy community to the general public and to policy and decision-makers.

This proposal sets out an implementation strategy for this Core Project Area. It aims to provide uniform access to a range of resources – data, models and software – by a range of users – scientists, policy-makers, students and the general public. It is based on the adoption and adaptation of Grid technologies, currently being developed to support e-science in other disciplines, to meet the specific requirements of canopy research and the deployment of research results in policy and education.

Mission

To provide integrated, web-based access to data, models and analysis and visualisation tools for canopy research, policy formulation and education.

Background and justification

Current ecological and environmental science is increasingly dominated by large, international research programmes. This is a welcome trend, bringing as it does a greater concentration of resources onto particular problems, and greater collaboration between researchers. However, these developments have not been matched by corresponding advances in the planning and integration of the research itself or of the results of the research: apart from some adoption of metadata standards, the way we organise the research, the way we handle data, and the way we develop integrative models has changed little in the last decade or so, despite landmark advances in internet technology, standards development, and modelling methodologies.

What is true of most ecological and environmental research is especially true for canopy research, with its particular problems of access, replicability and diversity of studies. Communication amongst research groups has been poor, with little international co-ordination of research or use of harmonised methods: Final Report, 2000. These problems are compounded by the restricted focus of most studies, the problems of scale and frame of reference, and the emphasis on descriptive rather than functional approaches. The report continues: "What is now needed is a broader vision and structures of communication that will facilitate international collaboration, well-planned studies and experiments, and inclusion of systematic dissemination of results to researchers in allied fields".

This proposal outlines just such a broader vision. It aims to put in place the database, modeling and communication foundations needed to maximise the effectiveness of canopy research. We of course recognise the fundamental necessity of high-quality science if the GCP is to meet its aims, along with conventional forms of communication between scientists, the general public, and those responsible for formulating policy. However, these by themselves are not sufficient: we must also ensure that scientific results can be integrated and delivered in the most effective way possible. This means making the most of recent advances in information technology, including developments in e-science, modeling, databases and collaborative working.

The central element of our vision is the development of a canopy research information strategy based on the internet: specifically, the Grid. ("The Grid" is the label for a suite of technologies for supporting e-science, and is currently receiving massive funding in the US and Europe: £120 million over 3 years in the UK alone.) The CanopyGrid will contain field data, derived results, models and metadata, plus software tools for information analysis, visualization and presentation. Users will be able to access information and software tools seamlessly, without regard for where the data originates or where the processing takes place.

Projected users of CanopyGrid will include the main stakeholders in the GCP. Research scientists will use CanopyGrid to assemble information from a variety of sources, to undertake analyses, to visualize complex data sets and to build and run models. Policy makers will use pre-built CanopyGrid applications to explore policy issues relating to climate change, biodiversity and local livelihoods. Local people will use it in a participatory manner, helping to formulate models of their local communities and envisioning future scenarios. CanopyGrid will have a major role to play in educating canopy researchers, students and the general public, by for example enabling them to enter a virtual canopy and to run simulation models.

The CanopyGrid will have a key role to play in relation to international Conventions, such as the Conventions on Biological Diversity and Climate Change. By providing uniform, objective access to diverse data sources, it will enable interested parties to demonstrate their meeting of their own obligations, or the failure of others to meet theirs. This will enable a much higher level of debate than can be achieved by printed documents and verbal exchange.

Goals
1. Maintenance of the canopy research reference database systems
Current maintenance and development work on the Big Canopy Database being developed at Evergreen State College (see previous chapter) will continue, and its use as a research reference and data repository will be promoted within the canopy research community. Crane site data will be integrated.

2. Development of a model library
Standard notation for representing models and model metadata will be developed, along with tools to enable researchers to 'publish' models in the same way that people currently publish html pages.

3. Development of tools for building and using models

A suite of tools for building and running models will be developed. Protocols to enable others to place their own tools onto the CanopyGrid will be prepared.

4. Development of a mechanism for linking data and models, and integration of 1, 2, 3

Common metadata notation for data sets and model input/output variables will be developed, to enable the rapid linking of data to models.

Workplan

The following workplan is expressed in terms of GCP Phases 1 to 3. The current proposal relates to a 2 year period 2002–2004, covering the latter part of Phase 1 and the start of Phase 2. The implementation of this workplan will depend on securing funds for the GCP and this project.

Phase 1: 2000–2002
Technology and literature review
Preliminary functional specification

Phase 2: 2002–2005
Deliver BCD as a Grid application
Integration of crane data into Big Canopy Database
Development of model library
Development of web-based delivery of models
Development of web-based linking of data and models

Phase 3: 2005–2010
The main activity during Phase 3 will be to migrate the demonstration system developed in Phase 2 to the Globus platform. Globus is being developed internationally as the standard platform for Grid-based systems. Currently, under development, and is a heavyweight product only suitable for use in well-resourced sciences such as particle physics and molecular biology. It is not at this stage suitable for deploying on the lowly PC of the typical canopy researcher, let alone students or policy-makers, but within 2 years 'Globus-lite' systems should be feasible.

This phase will also involve a shift towards a 'semantic web' approach, in which information is marked up with metadata which allows the meaning of data to be understood by computer programs. This requires, *inter alia*, a major effort by the research community to formulate standards, including the development of standard thesauri (along the lines of, say, the CABI thesaurus).

Outputs

Improved availability, analysis and integration of data as web databases will ensure greater access to the data, and the development of common metadata standards will make it easier to find particular types of data. Web-based access to analysis and visualisation tools will enable researchers to find and use such tools in a matter of minutes, without having to worry about downloading and installing software, compatibility issues, etc.

Improved linkage between data and models

It currently involves considerable effort to find data for particular models, or models capable of handling particular data. By developing common metadata standards for data and models, it will become far simpler to achieve this linkage.

More efficient, effective and rigorous modeling

With models being published in CanopyGrid, supported by a range of modelling tools, researchers will be able to construct models by simply downloading previously-developed submodels. They will also be able to deliver their models to the whole community at the press of a button, and will be able to understand other people's models much more easily than at present. Collaborative modelling will also become more common.

Improved communication within the research community

CanopyGrid will enable the web-based publishing of a range of resources, such as data, models, project information and software tools. This in itself will constitute a form of communication between researchers, and will be far more effective than at present. Perhaps more importantly, this will stimulate greater informal communication within the research community, as use of each others resources inevitably leads to interaction between the originators and users of these resources.

Improved delivery of research results to stakeholders

A key consequence of the CanopyGrid approach is that manual re-processing of information will be minimised. Tools will be developed for presenting research results in a form that policymakers can understand. Models developed for use by researchers can be wrapped up in a shell that makes them suitable for use by school children.

Reporting

The principal mechanism for informing others about CanopyGrid will be CanopyGrid itself. Documents describing the system will be published within it, and these will on the whole be closely integrated with CanopyGrid resources: instead of using static images, such as screen dumps, to illustrate a particular feature, readers will be taken directly to the resource itself.

Contact details

Dr R. Muetzelfeldt, University of Edinburgh, Darwin Building, The King's Buildings, Mayfield Road, Edinburgh EH9 3JU, United Kingdom. Email: r.muetzelfeldt@ed.ac.uk
Dr Judy Cushing, Lab I, The Evergreen State College, Olympia, 2700 Evergreen Parkway, Washington 98505, USA. Email: judyc@evergreen.edu

Literature Cited

Nadkarni, N M, *et al* (2000) *Forest Canopy Planning Workshop: Final Report,* The Evergreen State College. Unpublished Report, pp.25.

Studying the spatial complexity of forest canopies at variable scales with the TREESCAPE System

By
K R Halbritter

VEGETATION is a key item in the total set of landscape elements to be inventoried, classified, mapped and evaluated. Vegetation characteristics affect the vegetated and non-vegetated spaces of a landscape. Such relevant vegetation characteristics are determined by many variables. They can be characterized spatially (by dimension, morphology, structure, location, individual distribution etc), temporally (monotemporal, multitemporal; past, present, future) and at different scales of observation (e.g. stand, landscape, region). Vegetation assessment, inventory or monitoring by remote sensing are usually motivated mainly or entirely by either economic, ecologic or by combinations of both sorts of reasons.

Features crucial for classification and evaluation vary considerably. Their assessment is subject to considerable variation according to the client and issues pursued. Observation scale (stand-, landscape- or regional level observation) and numerous other factors that matter to influence a contractor's options for the choice of sensor(s) and technique(s). The client's issues are subject to numerous influences and thus variable.

The multi-use ecosystem inventory and monitoring system TREESCAPE makes a holistic contribution to assess the vegetation at great detail and with high complexity and completeness. It consists in novel methodologies and software tools. Its advantages over existing systems are its high degree of automation, scale

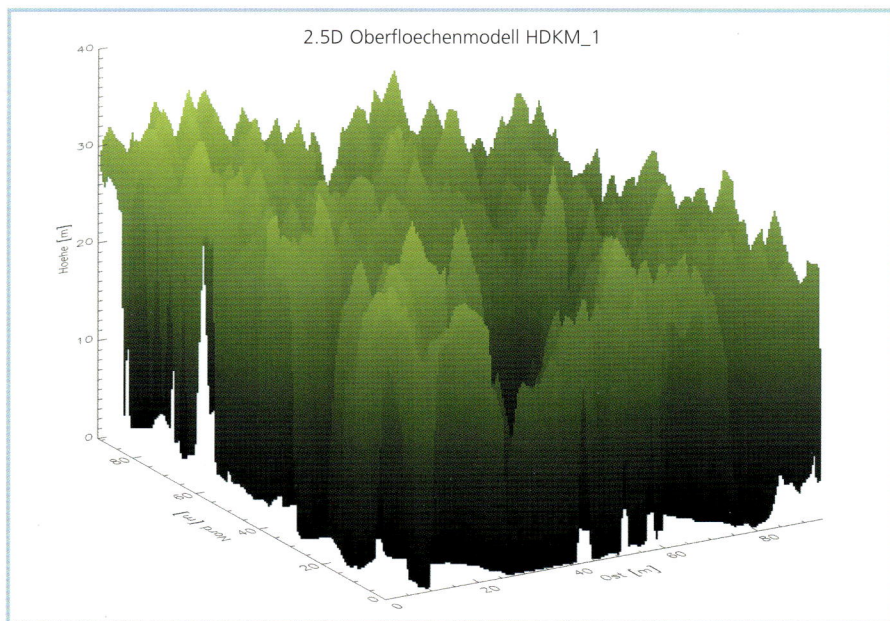

Figure 1: Shaded visualisation of a digital forest canopy surface model

invariance, strictly quantitative approach and elimination of atmospheric and terrain slope effects.

Among the TREESCAPE features are tools for remote sensing data enhancement and automated 1D-, 2D-, 2.5D- and 3D- data analysis, including automatic feature and object extraction, e.g.:

- visualisation of the canopy at high detail (still and animated pictures)
- localisation of plants as well as certain stand- and site features
- classification of plants, stand- and site features
- estimation of plant numbers and densities per area unit by classes
- estimation of biomass stock, increment and loss by classes
- quantifying diversities among individuals, communities, stands, landscapes, regions

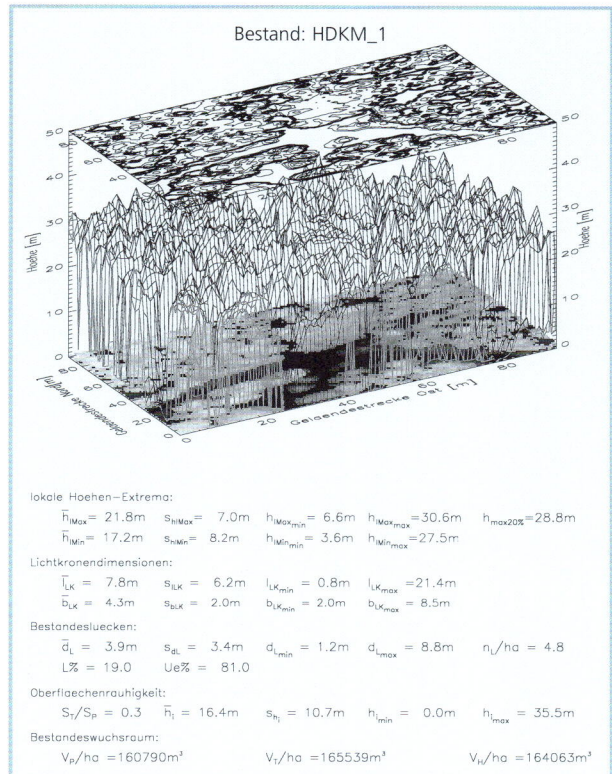

Figure 2: Surface model in three representations with statistics and indices. Top: Countour line model. Middle: Wireframe model. Bottom: Colour coded range image

- characterization of habitats and ecosystems according to pre-specified criteria
- provision of detailed data and information about the vegetation characteristics to be correlated with other data and information according to the client's needs
- provision of data with accurate georeference automatically
- automatic aggregation of results for any predefined scale of observation
- automatic generation of statistics for any predefined scale of observation
- options for further processing of TREESCAPE results with any other software
- integration of TREESCAPE results in any information system, with or without a geobase
- semi-automatic report generation of preselected TREESCAPE results using only alphanumeric results, only illustrations, only statistics or any combination of these

Figures 1–6 illustrate some of the performance of this new system for economic, ecological and integrated land information systems. They respresent only a fraction of the full set of alphanumeric and graphic results. This information was extracted automatically from a sample spatial data set of approximately one hectare. The georeferenced information can be point related, line related, area realted, volume related and time related information. The information relates to either the model surface, the model bottom or any invertal or volume between both boundary layers.

Thus, TREESCAPE serves any client no matter if he cares for detailed or overview knowledge on vegetated landscapes.

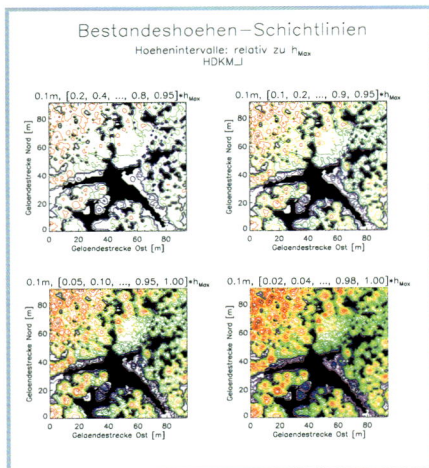

Figure 3: Contour line models of the canopy at different height intervals.

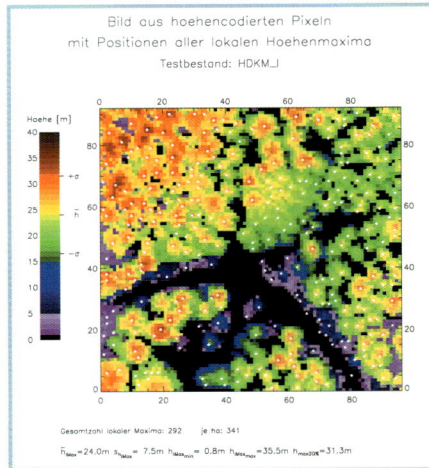

Figure 4: Colour coded range image using a special colour wedge. White dots: Tree positions. Below: Statistics and indices

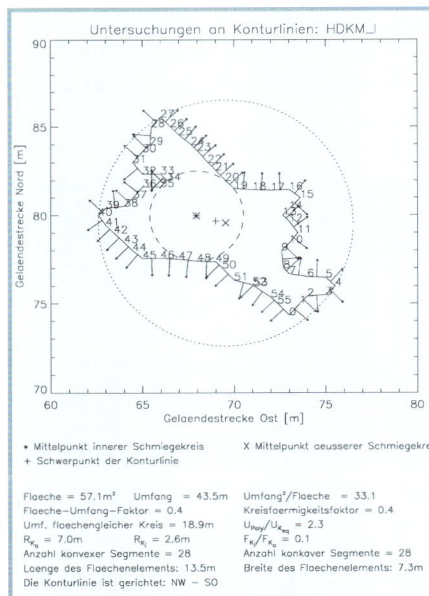

Figure 5 Single area element (contour) with statistics and indices This analysis is rather sophisticated and applies to contours of gaps, crowns, total canopy or other object contours

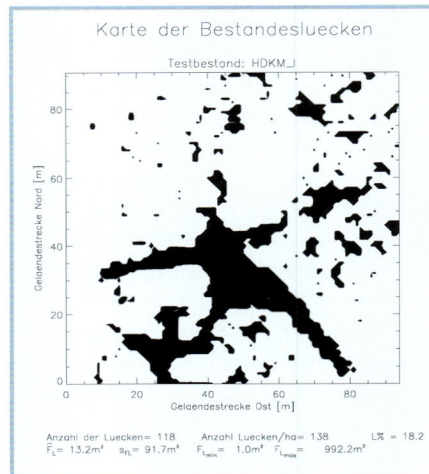

Figure 6: Map of canopy gaps with statistics, histogram indices.

Contact details

K.R. Halbritter, Droste-zu-Vischering-Weg 14, D-59227 Vorhelm GERMANY.
Email: khalbri@gwdg.de, Tel: +49(0)2528-1037, Fax: +49(0)2528-929496

Networks

12

Global Canopy Programme

By
Andrew W.
Mitchell

GCP
Global Canopy Programme *A global alliance linking studies of forest canopies worldwide into a collaborative programme of research, education and conservation addressing biodiversity, climate change and poverty alleviation. Presented here is the framework developed by the GCP Steering Committee up until the time of publication of this book. Details of individual projects can be found on the GCP website. Funding for Phase II and III of the Programme is currently being sought.*

Mission

The aim of the Global Canopy Programme (GCP) is to integrate forest studies across the world into a ten year linked program of research, conservation and education, focussed on understanding the critical role of forest canopies in biodiversity and climate change. It also aims to identify societal benefits from forest canopies, and transmit information to key stakeholders. This initiative evolved from an ESF/NSF funded International Canopy Science Workshop in Oxford, held in November 1999. At the workshop, a template for the GCP was produced by 29 international experts from 10 countries. They concluded that by working together, canopy researchers would be able to leverage more funding for a major collaborative Natural Science project to investigate "nature's last biotic frontier". They called for significant new funding on the scale of large Physical science projects (US$20-50 million) to undertake this pioneering task. To view the Workshop's Final report please see: http://192.211.16.13/individuals/nadkarnn/ESF.htm

The need

Such a programme is now urgently needed to plug major gaps in our knowledge. The structure, function and resilience of the world's forest canopy environment is unknown. Almost half of all terrestrial life forms could exist in forest canopies. A small fraction has been documented. The influence of forest canopies on climate change and their role in maintaining the earth's biological diversity is based on very limited knowledge. These roles are connected in forest canopies as nowhere else, through the forest's interface with the atmosphere. The Kyoto Protocol has focussed attention on the role of forest canopies in sequestering carbon from the atmosphere. Whether forests act as a sink or a source of carbon remains uncertain. How the mechanism works at the canopy atmosphere interface remains unclear. We do not know the economic value of canopy biodiversity or ecological services,

the value of its products for human health or its eco-tourism potential for local communities. The GCP will create a significant new international effort to throw light on these issues and will serve specific governmental requirements under the Conventions on Biological Diversity and Climate Change. Time is not on our side. The window of opportunity is closing as destruction of ancient forests proceeds apace worldwide.

Projects in development

Information will be made available on-line to the public, policy makers, and scientists, through the state of the art *Big Canopy Database* currently being developed with funds from the US National Science Foundation. A *Canopy Training School* will offer capacity building courses for scientists, forest managers and conservationists in biodiversity rich nations that need extra skills and will inspire new leadership in canopy science and conservation. A *CanopyLIFE* biodiversity rapid assessment programme is in development to assess the value of the myriad life forms in different forested environments. *FluxCAN* is an initiative to investigate the relationship between biodiversity and climate change and the process of carbon sequestration in the forest canopy. *CanopyWorld*, an interactive virtual rainforest website, is planned for schools. A number of 'Pathfinder Projects' on biodiversity and LIDAR laser scanning of canopy structure are already underway to demonstrate the comparative and collaborative capabilities of the GCP and the benefits they could bring.

GCP CORE PROPOSED PROJECT AREAS

ECOSYSTEM DYNAMICS

The ecosystem dynamics of forest canopies are critical to the functioning of the Earth's major life-support system, forests. This project series will investigate how forest canopies contribute to the overall dynamics of the forest ecosystem and will include research into structure/function relationships using state of the art technology.

BIODIVERSITY

Much of the world's biodiversity resides in the forest canopy and this programme will set out to investigate the world's last biological frontier. Much of this work will be on arthropods in the canopy. In particular issues related to biodiversity pattern, biodiversity process, the development of canopy based management tools and the evaluation of canopy-based ecological services will be investigated.

CLIMATE CHANGE

The forest canopy is where the land meets the atmosphere. The fluxes of gases which occur at this interface are thought to increasingly play a significant role in maintaining the Earth's climate and a number of studies suggest potentially very large values for the carbon storage functions of forests. The GCP will investigate the role forest canopies play in climate change and will collaborate with the major organisations tackling these issues.

SOCIETAL BENEFITS

The ecosystem function of forest canopies refers to the habitat, biological or system properties or processes of ecosystems. Ecosystem goods of forest canopies (such as food) and services such as interception of water represent the benefits human populations derive, directly or indirectly. The aim of this programme is to investigate and identify the benefits/social value that forest canopies can provide at the global level and at the local community level. This will include ecosystem services as well as economical benefits derived in the form of non-timber forest products.

CAPACITY BUILDING

The aim of this programme is to train host country nationals in the field of canopy science and conservation. Initially this will be in partnership with the Marie Selby Botanical Gardens in Florida and will eventually be expanded into a global programme. We want to help create the leaders for the future in this field, which is currently dominated by first world researchers. Canopy research facilities in the Pacific North West and in Brazil have also offered to assist as well as the Oxford Forestry Institute.

POLICY

Once comparative and collaborative research has been conducted this will provide the GCP with a much stronger tool to approach Governments and their global organisations. The GCP will be able to provide incentives and scientific support for decision makers for the justification of forest conservation due to the value of forest canopies.

COMMUNICATION

The GCP will encourage communication of information within the forest canopy community through ICAN and provide a means of communicating results to the wider public as well as policy and decision makers in a non-scientific form.

DATA MANAGEMENT

Through the Big Canopy Database and the proposed Canopy Grid modelling component, the GCP will foster Data Management to encourage data sharing and collaboration between canopy scientists around the world enabling larger, global, issues to be approached.

EDUCATION

With the data accumulated the GCP will foster the creation and dissemination of education, outreach and public relations materials to decision makers and the general public.

Key questions

The GCP will enable us to address some key questions significant to those outlined in the Stakeholder section below.

Ecosystem Function
- What is the relationship between structural diversity and biological diversity?
- What is the impact of natural disturbance on forest canopies?
- Does ecosystem complexity confer resistance or stability to disturbance?

Biodiversity
- What are the organisms that dwell in the canopy and which microhabitats do they occupy?
- What is the true extent of canopy biodiversity, and how is diversity maintained?
- What are the special requirements for canopy life and how are they threatened?

Climate Change
- What are the effects of canopies in taking up nutrients and CO_2, on albedo, on hydrological cycles?
- Can ecological effects at large scales be predicted from smaller ones?
- How will global change affect biodiversity held within the canopy?
- What is the forest canopy's role in carbon sequestration?

Societal Benefits
- What are the economic and social values of forest canopies?

- what are the effects of forest management practises on forest canopy biodiversity?
- how can canopy products and ecological services offer secure environment to local communities?
- can enhanced biological diversity improve the value of managed or plantation forests?

Organisational structures

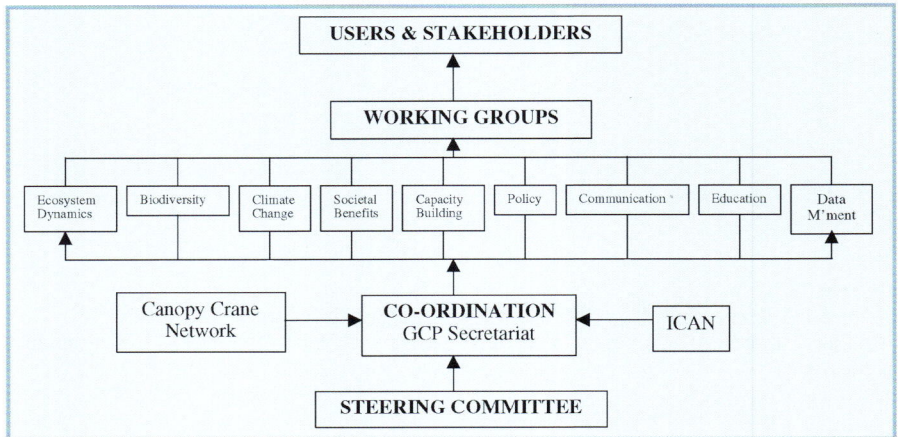

Principles

INTERNATIONAL LINKAGE: The GCP will work within the framework of conventions endorsed at the United Nations Conference on Environment and Development (UNCED), and in the spirit of The Rio Declaration.

BIODIVERSITY: The GCP will address issues and concerns raised by the Convention on Biological Diversity (CBD).

CLIMATE CHANGE: The GCP will address issues and concerns raised by the Framework Convention on Climate Change (UNFCCC).

RESEARCH AGENDA: The GCP will raise funds and disperse grants for collaborative research, and establish projects to carry out integrated interdisciplinary research between sites.

NEW TECHNOLOGIES: The GCP will help develop the innovative technological base for canopy research.

EDUCATION: Results from GCP will form the basis for an international education programme increasing awareness and understanding of the world's most biodiverse environment.

CONSERVATION: The GCP will actively promote the conservation and sustainable use of natural and managed forests.

NATIONAL AND INTERNATIONAL LINKAGE: The GCP will form links with organizations coordinating research relevant to the GCP and the International Conventions.

Stakeholders

The primary stakeholders and users for the global findings of the Global Canopy Programme will be the parties to forest related conventions – these include, Convention of Biological Diversity, Climate Change Convention and the UN Forum on Forests. Parties to these conventions will then determine which findings will be formally accepted into the individual convention process, based on their specific information needs.

Other important stakeholders include national governments, NGO's, Civil Society, Businesses, Local Communities, Museums, Media, Scientists, Forest Managers, Forest Ecologists, Schools and Educators. An advisory group will be established and the GCP will also establish links to national focal points for the Forest Related conventions in all nations.

Programme development

TECHNICAL EXPERTS: The Global Canopy Programme will be carried out through expert working groups focused on the CORE project areas. Each working group will be chaired by leading natural or social scientists from industrial and developing countries.

DESIGN and METHODS: The GCP will work with the latest technology in canopy access systems from satellites to balloons, construction cranes and walkways to climbing techniques. In its first phase, the Global Canopy Programme will focus on the development of an internally consistent set of methodologies for conducting the CORE projects at local, national, regional and global scales.

PEER REVIEW: All of the CORE projects will undergo peer review. Reviewers from a wide range of countries will be nominated by the Steering Committee and the Science Advisors. The review process will be developed and overseen by the Steering Committee.

Support for the exploratory phase 2000–2002

The Rufford Foundation   The Maurice Laing Foundation
The John Ellerman Foundation   European Science Foundation
US National Science Foundation

Time frame and outputs

The GCP envisages a ten year programme of research with significant scientific, education and conservation outputs as follows:

PHASE I: 2000–2002 Exploratory Phase

- identify stakeholders, collaborators and scope of work.
- establish effective working relationships amongst Forest Canopy Organisations and identify CORE projects.
- establish harmonised protocols for the conduct of projects.

- Implement pilot "Pathfinder Projects" demonstrating the effectiveness of coordination and collaboration in the CORE project areas.
- Launch GCP at the 3rd International Canopy Conference, June 2002

PHASE II: 2002–2005 Development phase

- Implement CORE projects at multiple sites including International Canopy Crane Network and begin data gathering process.
- Data co-ordination – Establish Canopy Grid to become a working service and encourage the use of it and the Big Canopy Database.
- Develop a state of the art multilingual web site for researchers as a source of information, decision makers and the general public – (including a "Canopy World" children's educational site) – to include web cams, links to relevant web sites.
- Review and redesign projects and identify locations for 8/10 new canopy cranes for Phase III.

PHASE III: 2005–2010 Full programme

- expand crane network – introducing 8/10 new cranes to obtain a more complete global representation.
- expand the use of collaborative research to be inputted into the Big Canopy Database.
- canopy Grid to become a commonplace tool amongst canopy researchers.
- expand developed projects to fully operational collaborative programmes at multiple sites.
- dissemination of education/outreach and public relations materials to decision makers and the general public.
- publication of findings and distribution of reports to stakeholders, scientific publications in peer review journals, exposure through television productions and the media.

Global Canopy Programme

Website: www.globalcanopy.org

Fundraising requirements

To date US$1,000,000 has been raised for Phase I of the GCP to form the Steering Committee and SAC, prioritise key science questions, define GCP CORE projects and identify lead institutions to carry them out. Phase II of the GCP is due for launch at the 3rd International Canopy Conference in Cairns, Australia on 28th June 2002. Phase II of the GCP is aiming to raise funding at the $5–6 million level for the period 2002–5. Phase III will aim for $25–30 million for the period 2005–2010.

The GCP Secretariat is hosted at the Global Canopy Foundation, based in Oxford, UK, Registered UK Charity (No. 1089110), Incorporated as a Company Limited by Guarantee (No. 4293417).

Contact details

Andrew Mitchell, Global Canopy Programme, Halifax House, University of Oxford, 6-8 South Parks Road, Oxford OX1 3UB, UK. Tel: +44(0)1865-271036; Fax: +44(0)1865-271035 Email: a.mitchell@globalcanopy.org

The International Canopy Network

By
Nalini Nadkarni
and
Joel Clement

Forest canopy communities are important in maintaining the diversity, resilience, and functioning of the ecosystems they inhabit. With the increasing interest in and amounts of data on forest canopies that are resulting from new canopy access techniques, ecologists require tools to communicate internationally, manage and analyze their data, and to compare data from disparate studies.

In 1993, with the support of the National Science Foundation, a group of canopy researchers initially established an organization to bring together forest canopy researchers, quantitative scientists, and computer scientists to develop methods to collect, store, display, analyze, and interpret three-dimensional spatial data relating to tree crowns and forest canopies. Their activities included:

1. Compiling the array of research questions and needs from canopy scientists via a survey to understand the characteristics of canopy scientists, especially their questions, pathways of communication, and the scientific issues that are now understood to or might potentially require information on canopy structure (Nadkarni & Parker 1994);
2. Examining potentially applicable information models and software tools that are in use in allied fields; and
3. Developing conceptual models for the types and format of information and analyses to answer research questions posed by forest canopy researchers.

They also built an interdisciplinary and international communication network. The International Canopy Network (ICAN) exists to facilitate communication among individuals and institutions concerned with research, education, and conservation of organisms and interactions in forest canopies (Nadkarni *et al* 1996). The institution was initiated as a non-profit organization in 1994, with headquarters at The Evergreen State College, in Olympia, Washington. Some of the current core activities of the ICAN include: facilitation of the electronic mail bulletin board (canopy@lternet.edu); production and circulation of the quarterly newsletter (titled 'What's Up?'), organization of scientific symposia and meetings, maintenance of a citations bibliographic database on scientific and popular aspects, distribution of a canopy researcher directory, and dissemination of information about forest canopies via our website. Over 750 forest canopy researchers for 62 countries now subscribe as members of ICAN to join see our website (www.evergreen.edu/ican)

Other activities are focused on outreach by scientists to segments of the general public. A major thrust is the creation of interpretive materials for school children by talks by canopy researchers in schools and publication of articles in kids' magazines. They also maintain the 'Ask Dr. Canopy!' program, whereby children who hear these talks or read our articles can write or email questions to 'Dr. Canopy', who is the collective persona of eight volunteer canopy researchers. Another area of outreach is to the media. The ICAN provides scientifically sound information to journalists, filmmakers, and television reporters who are interested in bringing the forest canopy to the awareness of their readers and viewers. More recently ICAN has played a major role in the development of the Global Canopy Programme.

The Board of Directors consists of eight members who represent the constituent fields of research, education, conservation, advocacy, and arboriculture. The Directors serve two-year renewable terms and meet annually. The Advisory Council, which consists of up to 20 members, takes part in decisions on an ad hoc basis.

ICAN is a self-supporting organization, funded by subscriber dues, donations, and grants. A regular subscribership costs US $30. This provides the following services: 1) four newsletters per year; 2) an annually updated directory of ICAN members (over 800 in 62 countries); and access to the bibliographical database at ICAN headquarters. Subscriber dues also support other ICAN activities in research, education, and conservation. Students may pay $US20 per year.

Contact details

For more information on the ICAN, contact Nalini Nadkarni
Dr Nalini M. Nadkarni, Lab II, The Evergreen State College, Olympia, WA 98505, USA
Email: nadkarnn@evergreen.edu
Joel P. Clement, Lab II, The Evergreen State College, Olympia, WA 98505, USA
Email: canopy@seanet.com
www.evergreen.edu/ican

Literature cited

Nadkarni, N. M. and G. G. Parker. 1994. A profile of forest canopy science and scientists – who we are, what we want to know, and obstacles we face: results of an international survey. *Selbyana* 15:38–50.

Nadkarni, N., G. G. Parker, E. D. Ford, J. B. Cushing, and C. Stallman. 1996. The International Canopy Network: A pathway for interdisciplinary exchange of scientific information on forest canopies. *Northwest Science* 70:104–108.

FLUXNET

FLUXNET is a global network of some 150 micrometeorological tower sites that use eddy covariance methods to measure the exchanges of carbon dioxide (CO_2), water vapor, and energy between terrestrial ecosystem and atmosphere. At present, over 150 tower sites are operating on a long-term and continuous basis. Researchers also collect data on site vegetation, soil, hydrologic, and meteorological characteristics at the tower sites.

The goals of Fluxnet are to understand the mechanisms controlling the exchanges of CO_2, water vapour and energy across a spectrum of time and space scales. Fluxnet also provides ground information for validating estimates of net primary productivity, evaporation, and energy absorption that are being generated by sensors on the Nasa Terra satellite.

Fluxnet also has a goal to provide information to Fluxnet investigators and the public. Fluxnet data available at the ORNL DAAC include monthly and annual heat, water vapor, and CO_2 flux, gap-filled flux products, ecological site data, and remote sensing products.

Data are available for the completed 3-year Euroflux project, which collected measurements of fluxes of carbon dioxide, water vapor, and energy exchange at 13 sites in Europe.

Fluxnet builds on regional networks of tower sites:
- South and North America (AmeriFlux);
- Europe (CarboEurope);
- Asia (AsiaFlux);
- Australia and New Zealand (OzFlux); and
- independent tower sites.

To-date there has been little opportunity for FLUXNET researchers to incorporate considerations of the impact of climatic or atmospheric change on biodiversity and the functioning of the ecosystem. This is a gap the Global Canopy Programme would like to bridge.

Website
www.daac.ornl.gov/fluxnet/

International Conventions relevant to Forest Canopies

By
Katherine Secoy

FOREST canopies are renowned for their great diversity and role in forest functioning, yet there are still great gaps in the understanding of this 'last biological frontier' – something that will be addressed by the GCP. Forest canopies are at the interface between forests and the atmosphere. They are where many of the processes take place important to forest-atmosphere inter-actions, such as photosynthesis, respiration, carbon flux, and water cycling. A number of international bodies and agreements are relevant to forest canopies. Research and conservation efforts can benefit from the policy context these provide in terms of fundraising and by using results to influence policy outcomes.

The Commission for Sustainable Development (CSD) was established in 1992 to ensure effective follow-up of UNCED (i.e. to implement Agenda 21), and to monitor and report on implementation of the agreements at the local, national, regional and international levels. The CSD established the Intergovernmental Panel on Forests (IPF) to address forestry issues within Agenda 21. The Interagency Task Force on Forests (ITFF) was formed to implement the IPF agenda. The ITFF have called for 'Scientific research, forest assessment, forest valuation, and development of Criteria and Indicators for sustainable forest management'.

There are three significant Inter-Governmental Conventions which can be related to Forest Canopies. These are the Convention on Biological Diversity, the United Nations Framework Convention on Climate Change and the Vienna Convention for the protection of the Ozone layer. Their significance lies in the fact that they set out the commitments for national and international measures aimed at tackling the current pressing global issues of biodiversity loss and climate change, related to forest ecosystems (amongst others).

Convention on biological diversity

In recognition that biological diversity is a global asset of tremendous value to present and future generations, the United Nations Environment Programme (UNEP) convened the Ad Hoc Working Group of Experts on Biological Diversity in November 1988 to explore the need for an international convention on biological diversity.

Work culminated on 22 May 1992 with the creation of the Convention on Biological Diversity. This was opened for signature on 5 June 1992 at the United Nations Conference on Environment and Development (the Rio 'Earth Summit'). To date there are 183 parties to the Convention of which 168 are signatories. The United Sates has yet to sign amongst others.

The overall aim of the Convention on Biological Diversity (CBD) is:
"To conserve biological diversity, promote the sustainable use of its components, and encourage equitable sharing of the benefits arising out of the utilization of genetic resources. Such equitable sharing includes appropriate access to genetic resources, as well as appropriate transfer of technology, taking into account existing rights over such resources and such technology."

CBD Article 25 called for the establishment of the Subsidiary Body for Scientific, Technical, and Technological Advice (SBSTTA), to provide the Conference of Parties (COP) advice relating to the implementation of this Convention. The GCP has worked to provide information relevant to the CBD via the SBSTTA. In November 2001, the SBSTTA met for the 8th time in Montreal, Canada to discuss specific issues relating to forest biological diversity. The GCP secretariat was successful in lobbying for greater emphasis in forest canopies. The SBSTTA recommendations for COP 6 included the following references calling for forest canopy research in the new work programme of the convention.

Programme Element 1, Goal 2, Objective 3, Activity a.
"*Promote monitoring and research on the impacts of climate change on forest biological diversity and investigate the interface between forest components and the atmosphere*"

Programme Element 3, Goal 3, Objective 1, Activity b
"*Develop and support research to understand critical thresholds of forest biological diversity loss and change, paying particular attention to endemic and threatened species and habitats including forest canopies*"

These were ratified at the COP meeting in the Hague in April 2002.

The significance of this is that for the first time the convention mentions forest canopies and the interface between forests and the atmosphere. Governments will need to report back at COP7 on progress they have made towards achieving these programme elements. This may help leverage funding for future canopy research and conservation efforts.

United Nations framework convention on climate change (UNFCCC) and its Kyoto Protocol
The United Nations Environment Programme (UNEP) established the Intergovernmental Panel on Climate Change (IPCC) in 1988. The role of the IPCC is to assess the scientific, technical and socio-economic information relevant for the understanding of the risk of human-induced climate change. The IPCC was instrumental in setting up the UNFCCC.

Framework convention on climate change (UNFCCC).
The Intergovernmental Negotiating Committee for a Framework Convention on Climate Change (INC) met for the first time in February 1991. After just 15 months, on 9 May 1992, the INC adopted the United Nations Framework Convention on Climate Change. The Convention was opened for signature at the UN Conference on Environment and Development (UNCED), the so-called "Earth Summit", in Rio de Janeiro, Brazil, on 4 June 1992, and came into force on 21 March 1994. Today, 186 governments and the European Community are Parties to the Convention. The United States has not signed yet. Parties meet regularly at the annual Conference of the Parties (COP) to review the implementation of the Convention and continue talks on how best to tackle climate change.

The Convention sets an 'ultimate objective' of stabilizing atmospheric concentrations of greenhouse gases at safe levels. Such levels, which the Convention does not quantify, should be achieved within a time frame sufficient to allow ecosystems